T0338608

Smart Grid Analytics for Sustainability and Urbanization

Zbigniew H. Gontar
SGH Warsaw School of Economics, Poland

A volume in the Advances
in Computer and Electrical
Engineering (ACEE) Book Series

Published in the United States of America by
  IGI Global
  Engineering Science Reference (an imprint of IGI Global)
  701 E. Chocolate Avenue
  Hershey PA, USA 17033
  Tel: 717-533-8845
  Fax: 717-533-8661
  E-mail: cust@igi-global.com
  Web site: http://www.igi-global.com

Copyright © 2018 by IGI Global. All rights reserved. No part of this publication may be reproduced, stored or distributed in any form or by any means, electronic or mechanical, including photocopying, without written permission from the publisher.
Product or company names used in this set are for identification purposes only. Inclusion of the names of the products or companies does not indicate a claim of ownership by IGI Global of the trademark or registered trademark.

Library of Congress Cataloging-in-Publication Data

Names: Gontar, Zbigniew H., 1964- editor.
Title: Smart grid analytics for sustainability and urbanization / Zbigniew H.
  Gontar, editor.
Description: Hershey, PA : Engineering Science Reference, [2018] | Includes
  bibliographical references and index.
Identifiers: LCCN 2017025680| ISBN 9781522539964 (hardcover) | ISBN
  9781522539971 (ebook)
Subjects: LCSH: Smart power grids. | Electric power consumption--Measurement.
  | Electric power--Conservation. | Sustainable urban development.
Classification: LCC TK3105 .S4868 2018 | DDC 621.319/2--dc23 LC record available at https://
lccn.loc.gov/2017025680

This book is published in the IGI Global book series Advances in Computer and Electrical Engineering (ACEE) (ISSN: 2327-039X; eISSN: 2327-0403)

British Cataloguing in Publication Data
A Cataloguing in Publication record for this book is available from the British Library.

All work contributed to this book is new, previously-unpublished material.
The views expressed in this book are those of the authors, but not necessarily of the publisher.

For electronic access to this publication, please contact: eresources@igi-global.com.

Advances in Computer and Electrical Engineering (ACEE) Book Series

ISSN:2327-039X
EISSN:2327-0403

Editor-in-Chief: Srikanta Patnaik, SOA University, India

MISSION

The fields of computer engineering and electrical engineering encompass a broad range of interdisciplinary topics allowing for expansive research developments across multiple fields. Research in these areas continues to develop and become increasingly important as computer and electrical systems have become an integral part of everyday life.

The **Advances in Computer and Electrical Engineering (ACEE) Book Series** aims to publish research on diverse topics pertaining to computer engineering and electrical engineering. **ACEE** encourages scholarly discourse on the latest applications, tools, and methodologies being implemented in the field for the design and development of computer and electrical systems.

COVERAGE

- Computer Architecture
- Chip Design
- Qualitative Methods
- VLSI Fabrication
- Microprocessor Design
- Analog Electronics
- Electrical Power Conversion
- Applied Electromagnetics
- Algorithms
- Digital Electronics

IGI Global is currently accepting manuscripts for publication within this series. To submit a proposal for a volume in this series, please contact our Acquisition Editors at Acquisitions@igi-global.com or visit: http://www.igi-global.com/publish/.

The Advances in Computer and Electrical Engineering (ACEE) Book Series (ISSN 2327-039X) is published by IGI Global, 701 E. Chocolate Avenue, Hershey, PA 17033-1240, USA, www.igi-global.com. This series is composed of titles available for purchase individually; each title is edited to be contextually exclusive from any other title within the series. For pricing and ordering information please visit http://www.igi-global.com/book-series/advances-computer-electrical-engineering/73675. Postmaster: Send all address changes to above address. ©© 2018 IGI Global. All rights, including translation in other languages reserved by the publisher. No part of this series may be reproduced or used in any form or by any means – graphics, electronic, or mechanical, including photocopying, recording, taping, or information and retrieval systems – without written permission from the publisher, except for non commercial, educational use, including classroom teaching purposes. The views expressed in this series are those of the authors, but not necessarily of IGI Global.

Titles in this Series

701 East Chocolate Avenue, Hershey, PA 17033, USA
Tel: 717-533-8845 x100 • Fax: 717-533-8661
E-Mail: cust@igi-global.com • www.igi-global.com

Table of Contents

Detailed Table of Contents

The term Industry 4.0 was born in the research group of the German federal government as well as in a project of the same name from the high-tech strategy of the federal government. It is meant to describe the interlacing of industrial production with modern information and communication technology. A key success factor and a major difference to computer-integrated manufacturing (according to Industry 3.0) is the use of internet technologies for communication between people, machines, and products. Cyber-physical systems and the internet of things (IoT) form the technological basis. The objectives are essentially the classic goals of the manufacturing industry, such as quality, cost and time efficiency, as well as resource efficiency, flexibility, convergence, and robustness (or resilience) in volatile markets. Industry 4.0 is one of the core themes of the federal government's digital agenda.

Chapter 2

This chapter is devoted to the study of the role of ecological subsystem in the structure of the sustainable development program of smart city. The author suggests the logic of building the environmental strategy of the city as a long-term landmark of its sustainable development including the environmental mission, vision of the future, goals and priorities, programs and their implementation, target indicators for assessing results, and consequences of realization programs. Certain attention is paid to the city as an object of research with a focus on environmental problems. The characteristics of the factors affecting the development of the ecological situation in the city are shown. A system of criteria and indicators that can be used to assess the impact of the planned environmental activities is proposed.

Chapter 3

Marisa Analía Sanchez, Universidad Nacional del Sur, Argentina

Organizations are experiencing a transformation as a consequence of digital technologies such as social, mobile, big data, cloud computing, and internet of things. The transformation presents challenges at several levels, and project management is not an exception. There are changes in the project environment, the power structures, capabilities, skills, and standard practices, just to name a few. Considering the eventual obsolescence of many project portfolio management practices, the aim of this chapter is to discuss the influence of internet of things in this discipline. The analysis departs from rethinking project management insights and describes the impact of smart and connected products considering many dimensions. Recommendations for each PPM stage are developed, followed by a brief discussion of future research directions.

Chapter 4

Shaun Joseph Smyth, Ulster University, UK
Kevin Curran, Ulster University, UK
Nigel McKelvey, Letterkenny Institute of Technology, Ireland

The introduction of the 21st century has experienced a growing trend in the number of people who choose to live within a city. Rapid urbanisation however, comes a variety of issues which are technical, social, physical and organisational in nature because of the complex gathering of large population numbers in such a spatially limited area. This rapid growth in population presents new challenges for the already stretched city services and infrastructure as they are faced with the problems of finding smarter methods to deal with issues including: traffic congestion, waste management and increased energy usage. This chapter examines the phenomenon of smart cities, their many definitions, their ability to alleviate the discomforts cities suffer due to rapid urbanisation and ultimately offer an improved and more sustainable lives for the city's citizens. This chapter also highlights the benefits of smart grids, their bi-directional real-time communication ability, and their other qualities.

Chapter 5

Luke Amadi, University of Port Harcourt, Nigeria
Prince I. Igwe, University of Port Harcourt, Nigeria

Since the 1990s, the field of smart grid has attempted to remedy some of the core development deficiencies associated with power supply in the smart city. While it seemingly succeeds in provision of electricity, it fails to fully resolve the difficulties associated with sustainable energy consumption. This suggests that the future of smart grid analytics in the smart city largely depends on efficiency in energy consumption which integrates sustainability in the overall energy use. This chapter analyzes the nexus between smart grid, sustainable energy consumption, and the smart city.

Chapter 6

Arash Anzalchi, Florida International University, USA
Aditya Sundararajan, Florida International University, USA
Longfei Wei, Florida International University, USA
Amir Moghadasi, Florida International University, USA
Arif Sarwat, Florida International University, USA

The rapid growth of new technologies in power systems requires real-time monitoring and control of bidirectional data communication and electric power flow. Cloud computing has centralized architecture and is not scalable towards the emerging internet of things (IoT) landscape of the grid. Further, under large-scale integration of renewables, this framework could be bogged down by congestion, latency, and subsequently poor quality of service (QoS). This calls for a distributed architecture called fog computing, which imbibes both clouds as well as the end-devices to collect, process, and act upon the data locally at the edge for low latency applications prior to forwarding them to the cloud for more complex operations. Fog computing offers high performance and interoperability, better scalability and visibility, and greater availability in comparison to a grid relying only on the cloud. In this chapter, a prospective research roadmap, future challenges, and opportunities to apply fog computing on smart grid systems is presented.

Chapter 7

Metering side of electricity distribution system has been one of the prime focus of industry and academia both. The most recent advancement in this field is installation of smart meters. The installation of smart meters enables collection of massive amounts of data regarding electricity generation and consumption. The analysis of this data could help generate actionable insights for the supply side and provide the consumers demand management-related inputs. The problem addressed in this chapter is to identify suitable data mining algorithm for applications like: estimating the demand and supply of electricity, user and use profiling of commercial, and industrial customers, and variables suitable for these purposes. This chapter, on the basis of rigorous literature review, presents a taxonomy of smart meter data mining. It includes the summary of application of smart meter data analytics, characteristics of dataset used, and smart meter business globally. This chapter could help researchers identify potential research opportunities, and practitioners can use it for planning and designing a smart electricity system.

Smart grid is a modern power grid infrastructure for improved efficiency, reliability, and safety, with smooth integration of renewable and alternative energy sources, through automated control and modern communications technologies. The smart grid offers several advantages over traditional power grids such as reduced operational costs and opening new markets to utility providers, direct communication with customer premises through advanced metering infrastructure, self-healing in case of power drops or outage, providing security against several types of attacks, and preserving power quality by increasing link quality. Typically, a heterogeneous set of networking technologies is found in the smart grid. In this chapter, smart grid communications technologies along with their advantages and disadvantages are explained. Moreover, research challenges and open research issues are provided.

Chapter 9
Understanding Smart City Solutions in Turkish Cities From the Perspective

H. Filiz Alkan Meshur, Selcuk University, Turkey

The purpose of this chapter is to analyze the concept of smart city and its potential solutions to correct urban problems. Smart city practices and solutions have been investigated through the lens of a sustainable perspective. As the general practices in the global scale were examined, particular focus has been directed to smart city practices in Turkey and applicable suggestions have been developed. A number of cities in Turkey rank the lowest in the list of livable cities index. Consequential to the rapidly rising population ratios, the quality of provided services declines; economic and social life in cities are adversely affected and brand images of cities are deteriorated. With the implementation of smart city practices, such problems could be corrected, and these cities could gain competitive advantage over their rivals. The key component of this smart administration is to most effectively utilize information and communication technologies during each single step of this process.

Preface

INTRODUCTION

With an abundance of smart grid analytics strategic options it can be appealing to search for transformation they could bring to urban processes management in smart city. The idea of a smart city is part of the concept of the "third wave" created by Alvin Toffler (1980), in which he proposed a new model of economics based on smart society, smart analytics, sustainable development goals and ICT infrastructure (now: internet of everything, big data, cyber-physical systems). Toefflers in the book *Creating of a new civilization: Third wave policy* (Toffler & Toffler, 1995) proposed a complete change of the classic concept of industrial production, a new lifestyle, changes in the way of work, referring to life, a new shape of economic life, new consciousness. Although smart grid analytics technology have matured and become available to the utility corporations, it is implemented by smart cities mainly as living laboratories. Smart cities themselves can even be considered as living laboratories of third wave concept. With this in mind, we propose an alternative method to enhance smart grid analytics activities in smart cities, starting with strategy - in the 21st century it is the new paradigm of industry 4.0 in association with urban sustainability - and operationalizing the smart city strategy throughout the smart grid analytics initiatives managed by new form of project management that incorporate sustainability issues with smart urban environment. The purpose of this book is to present this idea and up-to-date survey of applications in smart grid analytics that could be incorporated in smart cities to solve urban problems.

ORGANIZATION OF THE BOOK

We organize the book into nine chapters, that represents major topics in smart grid analytics and its implementation in smart cities.

Chapter 1, "Industry 4.0: Nothing Is More Steady Than Change," presents main strategic concerns of smart cities in the sense of reindustrialization. The main goal of the Industry 4.0 revolution is the creation of a smart industrial grid and the dissemination of grid technologies. The concept of grid, presented in the book, goes beyond the standard understanding of it. Traditionally, grid means the power grid. In computing, the concept of grid was defined by Ian Foster et al. (2001). It means creating a connection between computing resources of computer centers in order to create a virtual supercomputer, whose computing resources exceed the computational capabilities of each computer center separately, and which provides and shares computing resources of all computing centers included in the grid. Thus, it is possible to remotely carry out calculations in computer centers in various parts of the grid. Grid user does not have to have specialized knowledge and know complicated procedures in the field of grid management systems. The grid itself determines what computational resources will be needed for a given task and which resources will be the fastest available. The concept of Industries 4.0 assumes introducing the idea of SOA[1] for industrial automation. All control functions in the control system of an industrial enterprise must then be organized as services. At the higher levels of the automated control pyramid there are only SOA-IT software components. At lower levels, services are no longer just software functions, but they also represent mechatronic functions for the implementation of real processes that can affect the physical state of mechatronic elements controlling technical objects. It means the need to integrate Enterprise Resource Planning (ERP) software, Manufacturing Execution System (MES) and System Control and Data Acquisition (SCADA), control systems and energy management systems. The goal is to achieve high energy efficiency, which requires the cooperation of different applications. Therefore, the urban industry environment in industry 4.0 will create an industrial grid. A smart urban industry grid will enable the use of these resources to create integrated production processes for enterprises and creating a virtual organization. The industrial grid concept refers to known theories about a networked production system. For example, the concept of Disruptive Network Approach (DNA) is used by StreetScooter GmbH, which was founded at the University of Aachen to develop an electric car[2]. A smart industrial grid will enable sharing of production resources. This target system will be reached in stages, the first of which is the idea of an industrial cluster. The industrial grid will be geographically dispersed, heterogeneous in the sense of production and process resources - possessed and managed by various organizations - dynamic in the sense of variable availability over time, connected by a heterogeneous power and digital network. The features of an industrial grid will not differ from the characteristics of an IT grid system, namely: the autonomy of production resources will be maintained in the sense of local control over resources and local access policies to resources, otherwise: resources will not be managed centrally. The aim will be to optimize the energy efficiency of the production process. The main assumption of the industrial grid will be to separate the tasks of the business process into individual threads. An enterprise using a grid system will not need to know where the production resources will be taken from, where the enterprise integrated in the cluster / grid will carry out its tasks and which parts of the business infrastructure will be involved in its implementation. The city that decides to be the first to introduce a live industrial grid laboratory will gain an advantage not only in the field of reindustrialisation, but also in the field of sustainable development on the principle of "first takes everything".

Grid technologies in the power industry will allow the creation of powerful virtual power plants from a huge number of combined, heterogeneous micro-plants using renewable energy sources. In management, however, they will allow the creation of enterprises that could compete with large corporations. Anyone on the power grid will be able to be a micro grid in the smart grid. Each organizational unit (enterprise or institution) can be both a producer of energy and its consumer. Analogous policy towards enterprises will allow the creation of powerful virtual production resources from a huge number of combined, heterogeneous production micro-resources that provide their production resources for tasks of business processes that require more resources than those available to enterprises individually. Obviously, grid technologies and intelligent organizations based on them will affect management concepts.

Under the conditions of the Industry 4.0 revolution, each company will become an active node of the smart grid, which undertakes activities in the field of energy generation, production, transmission and distribution. For the purpose of determining the organization that plays the dual role of a producer and consumer of electricity, the prosumer concept is used.

The above issues determine the need to implement smart analyzes in cities related to the energy efficiency of business processes. The new role of the city in the smart grid, as a producer and supplier of energy, and the possibility of using the power grid and the digital network to create an industrial grid, give us a closer look at the methods of analysis related to energy management (Smart Grid Analytics) in energy industry and business process analysis (Process Analytics) and methods of their implementation. Regardless of the scale of considerations: a single organizational unit or a network of units organized into a cluster or grid, we will have to deal with a structure composed of small self-regulating systems in the energetic sense. The assessment of organizational units will be determined by their ability to adaptively react to the impact of the environment in terms of energy efficiency. A single prosumer, prosumer cluster or industrial grid will be treated as a smart organization - and it will be expected to constantly find and use success factors related to energy efficiency.

A smart organization (enterprise and institution) is understood in this book as a single prosumer (enterprise or institution), as a cluster of enterprises (or institutions) in the sense of an energy micro-network or as an industrial grid. Creating clusters of smart enterprises or industrial grid should be understood in the sense of structuring a smart organization. Structuring the energy efficiency of business processes should be the structuring factor of a smart enterprise. The basic objective of this analysis would be to investigate whether the creation of a cluster of enterprises or industrial grid would affect the technical and economic optimization of smart organization: whether existing business processes can be more energy efficient (or business processes can be implemented more effectively in the energy sense).

Chapter 2, "Conceptual Foundations of Creating Sustainable Development Strategy of Smart Cities: Environmental Aspect," discusses forms of implementing sustainable development in smart and friendly city. The aim of the chapter is to review current researches on sustainable development strategy. The structure of the review has been developed on the basis of a conceptual framework that unite strategic management, project management, and sustainable development.

Chapter 3, "How Internet of Things Is Transforming Project Management," describes a new approach for project management in smart environment.

Chapter 4, "Smart Cities, Smart Grids, and Smart Grid Analytics: How To Solve an Urban Problem," illustrates how to incorporate smart grid analytics in everyday life of smart cities. Smart grid analytics is the basic building block to implement smart city concept in the future. Smart grid analytics allows to be get acquainted with issues concerning the effectiveness and efficiency of smart city's processes in the sense of energy usage. The question is: how smart grid analytics could help smart cities in sustainable strategy. The assumption is that it would be connected with processes' analysis. The first stage is a statistical analysis of completed off-line processes. It can be viewed as process control or process audit. Analyzed processes do not have to be documented. Such a model can be inductively generated by the process mining algorithms. It is possible to create a model of the flow of control in the process (ordered sequence of the events in the process), social network model (relationship between the actors of the process) or the organizational model (structure of the units jointly implement the process). The control flow model usually takes the form of a graph (e.g. Petri net, BPMN, EPC, or UML activity diagram). The collected event data can be stored in any way (e.g. In the SAP Event data are written to the databases). Some systems for processes exploration use dedicated formats (MXML, XES), which requires data analysts additional obligation associated with the transformation of flat files to this format. Statistical analysis of the process applies key performance indicators of processes, the energy usage of each activity and the overall process, and analysis of the functioning of the process (discovery rules for the functioning of processes). Built process model can be used to create simulation models. It is also possible to expand, modify and optimize the model. The second stage includes the analysis of the current processes in the sense of their implementation, in real time. Of course, the process can be characterized by the path of the process and actors of the process. However, the process may also be characterized by the values of event data, and show efficiency of energy usage. The second stage allows the examination of compliance (monitoring of deviations of the process from the assumed model). Event data must be monitored in real time to feed the analytical module. Perhaps, in the near future event data will be stored in one place - in the event log - in a standard format. The current proposal is the standard Business Process Analytics Format (BPAF). The third stage of the analysis is a prospective analysis enables determination and assessment of variants of the process and the selection of optimal variant. It is predictive analysis and recommendations are generated in real time. When events have timestamps, then it is possible to discover bottlenecks, measure service level monitoring and to predict the total usage of energy of the current process. So, in the future smart grid analytics may be close connected with process mining. The chapter is the results of the research on the state-of-the-art of current understanding of smart grid analytics.

Chapter 5, "Analysis of the Nexus Between Smart Grid, Sustainable Energy Consumption, and the Smart City," is an interesting view on the problems of integrating smart environment with sustainable paradigms. The current models do not explain the behavior of the cities under the conditions of the smart power grid. It is possible to define ad-hoc model based on the assumption that the smart power grid will enable the triumph of economic co-operation (other concepts of shared economy, mesh economy, collaborative economy) for sharing energy electricity and production resources. A smart city with smart electric grid can be called a virtual cluster, which, unlike a real cluster, is not limited by proximity. The concept of a virtual cluster is easy to understand for IT professionals who use the term "virtual machine". Well, computer programs do not need to work directly with computer hardware, but with a management program (otherwise - the operating system). A user working on a computer does not refer to hardware, and to an operating system that specifies the use of hardware. It is said that the user has to deal with a "virtual machine", a computer model, and not a computer. Similarly, in a virtual cluster, enterprises do not have to collaborate directly with other enterprise resources, but only with a management center (Smart City Competency Center). An enterprise seeking to utilize virtual cluster production resources does not have to collaborate with a resource-owning enterprise. Determines the possibilities of using production resources. The concept of a competence center as an integrator of a smart city was proposed in many studies. The proposed activities for the competence center relate to energy management, which focuses on the energy efficiency of processes and applies to any city and business process management. The following areas of activity in the aforementioned areas seem to crucial: the forecasting of electricity demand for the processes of smart city and the assessment of the economic potential of radical innovation initiatives related to smart cities.

Chapter 6, "Future Directions to the Application of Distributed Fog Computing in Smart Grid Systems," refers to the advances in industrial embedded analytics systems and the greater reliance of cities on cloud computing, fog computing and mist computing.

Chapter 7, "Mining Smart Meter Data: Opportunities and Challenges," refers to the use of computational intelligence (CI) for problems, that in the past were strictly connected with short term load forecasting (STLF), and the short-term planning of energy market transactions. Computational intelligence deals with theories and methods of problem solving that are not efficiently algorithmized and which require a particular approach. These approach could be the basis for further research into the construction of a city model under smart grid conditions. The concept of the smart power grid was introduced in 2005 by S. Massin Amin and Bruce F. Wollenberg in the article "Toward a Smart Grid" for the design of a power system built around the idea of converting passive energy consumers to active power grid nodes, taking action on energy production and management (reducing consumption, affecting energy consumption by controlling high power consumption, etc.), supplying renewable energy (photovoltaic, wind power, small hydropower plants, cogeneration) and energy storage. According to the idea of smart power grid, every consumer of energy, after installing the necessary information and communication technologies (ICTs), becomes a producer and distributor of energy. The prosument concept was coined by Alvin Toffler in his book "The Third Wave" and developed by Don Tapscot in "The Digital Economy: Promise and Peril in The Age of Networked Intelligence". The similarity of problems in cities considered as active nodes of the smart power grid to the problems of existing energy market players (energy utility companies) makes it possible to consider its implications for forms of energy management analysis. The goal is to achieve efficiency of processes or additive value of equipment or systems that would not be achievable without smart grid analysis. This is the result of a deliberate strategy, planning, and programs, and this requires human intelligence and a smart society. This gives the opportunity to build models of social systems and models of technological systems that are smart: smart society, smart city, smart enterprise, smart grid, smart appliances industry, smart economy, smart mobility, smart democracy. We are dealing here with a chain of artifacts with a smart attribute, characterized by a multitude of connections and interactions, resulting in the construction of a new social system with a smart attribute. The origins of this brainstorm are connected with smart grid. Smart refers to new technical devices and to social systems. They create a causal chain in which individual cells can be isolated and subjected to separate tests. Smart meters installed in the smart grid force smart home appliances, smart buildings, smart grid computing, smart grid competency center that can be used for research and analysis - smart grid analytics for a demand-driven approach to purchasing energy, changing the concept of smart city vision, or smart grid society.

Chapter 8, "Communications Technologies for Smart Grid Applications: A Review of Advances and Challenges," refers to the advances in information and communication technologies and the greater reliance of cities on data analysis.

Chapter 9, "Understanding Smart City Solutions in Turkish Cities from the Perspective of Sustainability," concludes the considerations given in the book with an overview of case study.

Zbigniew Gontar
SGH Warsaw School of Economics, Poland

REFERENCES

Foster, I., Kesselman, C., & Tuecke, S. (2001). The Anatomy of Grid: Enabling Scalable Virtual Organizations. *International Journal of Supercomputer Applications, 15*(3).

Toffler, A. (1980). *The Third Wave*. Bantam Books.

Toffler, A., & Toffler, H. (1995). *Creating A New Civilization: The Politics Of The Third Wave*. Turner Publishing.

ENDNOTES

[1]   Service-oriented architecture (SOA) is an abstract concept of software architecture representing various methods or applications as repeatedly used and open services, which enable multiple implementation independent of the type of platform.

[2]   In 2014 Deutsche Post DHL Group was purchasing StreetScooter company, and since then exploit StreetScooter electric delivery vans.

Introduction

Smart cities continue to deploy digital infrastructure for more sustainable urban systems. Since early 21st century, smart grid analytics gives opportunities to take advantage for sustainable benefits. It's more than just providing opportunities to use smart grid data to improve urban processes. Smart grid analytics infrastructure is a software-defined platform for a strategic planning approach to improve urban life.

Unfortunately, there have been no comprehensive books on smart grid analytics technologies as feasible alternative to improve urban sustainability. This book presents important aspects of urban strategic planning. One of the most critical issues for the 21st century is the new paradigm of industry 4.0 in association with urban sustainability. The book presents brief discussion of the state of the art in Industry 4.0 from inside German perspective. In addition, the book examines Siberian mechanism about sustainable urban environment, and international view of development in design of smart grid analytics facilities, programmes, and polices. Finally, the book explores current state of project management in smart environment. To operationalize the smart city strategy throughout the smart grid analytics initiatives, new sustainable project management ways of working are implemented.

The book can be used as reference book which offers interesting insight into smart grid analytics issues in smart and sustainable cities.

Zbigniew Gontar
SGH Warsaw School of Economics, Poland

Chapter 1
Industry 4.0:
Nothing Is More Steady Than Change

Torsten Zimmermann
NUCIDA GmbH, Germany

ABSTRACT

The term Industry 4.0 was born in the research group of the German federal government as well as in a project of the same name from the high-tech strategy of the federal government. It is meant to describe the interlacing of industrial production with modern information and communication technology. A key success factor and a major difference to computer-integrated manufacturing (according to Industry 3.0) is the use of internet technologies for communication between people, machines, and products. Cyber-physical systems and the internet of things (IoT) form the technological basis. The objectives are essentially the classic goals of the manufacturing industry, such as quality, cost and time efficiency, as well as resource efficiency, flexibility, convergence, and robustness (or resilience) in volatile markets. Industry 4.0 is one of the core themes of the federal government's digital agenda.

ECONOMIC DEVELOPMENT SINCE 1990

An old proverb says: Nothing is more steady than change. And in our present day, this sentence seems to be more valid than ever: The East-West confrontation, which dominated the world after the Second World War, has been replaced by a political and above all economic globalization. Across the globe, this fundamental change has led nations, economies, and organizations to adapt their own strategies and orientations according to the changes. This gave rise to new forces, which also dramatically influenced cooperation between the individual partners. For decades,

DOI: 10.4018/978-1-5225-3996-4.ch001

Copyright © 2018, IGI Global. Copying or distributing in print or electronic forms without written permission of IGI Global is prohibited.

Figure 1.

the political and economic leadership role in the Western world was considered a given. Today, however, this is anything but self-evident. The Third World, marked by industrializations in the 1970s, metamorphosed into a political and economic partner of global importance. The individual economic regions then changed very differently. In this respect, a graphic by the VDMA / OECD of January 2013 is very revealing (VDMA, n.d.): According to it, Germany and Austria are the only leading industrial nations in the Western world who have been able to maintain the share of the manufacturing sector over the past 20 years. Since the 90's of the last century, Germany has been confirming its share in the producing sector between 25% and 30% in the gross national product (VDMA, n.d.). Austria is just behind with 23% to 25%. This is particularly noteworthy as the economic slump of the two countries triggered by the banking crisis in 2008 had already been overcome by 2011. The only other country to achieve this was the United States of America, albeit at a much lower level. All other industrial nations, however, are now levelling in on the 10% mark. Internationally, this trend motivated abolishing the manufacturing industry. Many industrial nations, therefore, depended on financial services and production took place where it was more affordable. Industrial equipment, expertise, and employees were therefore reduced. In Germany, on the other hand, the manufacturing sector remained a very important economic factor. The German industry kept the production largely in its own country and stepped up to automation in response to increased wage and production costs. Here, too, it led to a comprehensive structural change with economic and social policy aspects. In the course, some industries have partially disappeared (examples: textile industry, entertainment electronics, parts of shipbuilding) while others have emerged (examples: automotive, aircraft construction, medical technology, machine and plant engineering). The demand for German products from the latter areas is stronger than ever. The key factors in this are product quality, reliability and innovative functionalities, which often determine that the competition from other countries is not selected by the market. So the price plays a secondary role. In addition, high-quality, product-related services support user acceptance and customer satisfaction for German quality products.

How did the German industry manage to achieve this? The following key points should be mentioned:

1. High level of security and robustness in the development and production processes represents the basis for the production of high-quality products.
2. Comprehensive automation with high process quality allows a production with high unit numbers, whereby also product variants are supported.
3. Extremely effective use of new technologies (in particular IT tools for engineering, product validation, production planning, commissioning and production), which clearly distinguishes the German industry from the competition.
4. Skilful software embedding in all product types and categories (embedded systems), which often makes the world's most coveted innovative functions possible.

These four factors lead to the question of how intelligent production should evolve as a further development of the fields product engineering, process engineering, product validation, product planning, production planning, etc., for the next decades. This decade, this was the beginning of the new discipline Industry 4.0. To this end — driven by the German industry — a number of working groups have emerged, which determine I 4.0 solutions. Since the holistic approach (horizontal and vertical integration) is imperative, cooperation across company boundaries is necessary.

It is also important to quickly exchange data with various organizational units, sometimes including many and complex data objects. For with regard to the definition of Industry 4.0 as described at the outset, the information, consisting of accurate data, is the success factor for Industry 4.0. This is why the term "Big Data" or "data-driven business" often occurs in connection with Industry 4.0. Whether this topic must always occur in combination with Industry 4.0 is to be questioned. There are already some I 4.0 examples, which do without Big Data. However, Big Data solutions are used for most field solutions.

Despite the many research projects and working groups on this subject area, implementation into a wider range has been difficult. The applications are, for example, restricted to partial areas of a production line or small organizational units. The horizontal and vertical integration, required intelligence as well as flexible production and scalability represent high challenges in the context of concrete implementation.

The Fourth Industrial Revolution

The term "Industrial Revolution" refers to the changes that have taken place since the middle of the eighteenth century. The professional world distinguishes four phases. This is how the name Industry 4.0 came about for the latest changes within the industry, coined, for example, by the keywords "Internet of Things", "Cyber Physical Systems" and "Big Data". However, it was by no means the case that the Industry x.0 approach was coined when the steam engine was first introduced. Rather, the terms Industry 1.0 to 3.0 have only been used since 1970, at the beginning of the CAD/CAM era. Industry 4.0 is often referred to as an industrial revolution. Perhaps this is because industry-based business models often have a disruptive character. Certainly, the partially dramatic technological upheavals within the production of goods have also contributed to this. In my estimation, however, the evolutionary approach prevails, as the transformations can continue for decades. Many important elements are not even being considered yet. Not to mention the appropriate solutions.

A further discussion point is whether Industry 4.0 is driven by technology or by the market. It is certainly true that without the advances in technology, such as the miniaturization and performance of electronic components, as well as the internet as the communication medium, Industry 4.0 would be unthinkable. However, I consider the market as the more important driver compared to technology. Without significant demand, the transformation to Industry 4.0 makes little sense for a company. This observation might seem provocative. My experience with I 4.0 projects shows, however, that these — in spite of all technological advances — can definitely end in failure if the customer benefit has not been fleshed out comprehensively. Fortunately, the positive examples prevail, offering encouragement. These projects are distinguished by a high proportion of individual implementations, taking into account the specific environment. In addition, all successful transformations have in common that the customer benefits have been framed excellently. Some of them will be introduced in the next episode of this series. In this issue, however, I would first like to address the challenges.

The Challenge of a Holistic Approach

An often underestimated aspect is the need for a holistic approach when one wants to transform an organization to Industry 4.0. This means that a horizontal and vertical integration of all processes and organizational units must take place. If this is not taken into account, synergies within the company's own organization as well as with its business partners, the unique features of its products, and services and additional benefits for customers cannot be developed or realized extensively. However, especially in the age of globalization, these aspects are becoming increasingly important. Since in the truest sense of the word every stone needs to be turned around and questioned, this often leads to rebuilding organizations, structures, business rules, and processes. I already mentioned the term "disruptive business model". The new functionalities that make intelligent, networked systems possible can shift the focus of the company's offer from product to service. In any case, the importance of services will increase compared to today. Within the framework of the company strategy, the management must determine the importance of products compared to services within the company portfolio. Services could be product add-ons, elements of a service portfolio that equally exists alongside the product portfolio, or positioned as central solutions with downstream products. The strategic decision should be made on the basis of the prevailing market contexts and expectation horizons in the relevant market segment. I mentioned the synergy concept. Is it about cost and time advantages? Possibly, but not only: Improvements in flexibility, customer loyalty, transparency, quality, forecasting and work identification often represent the motivation for transformation according to I 4.0. When we say service, we often

Figure 2.

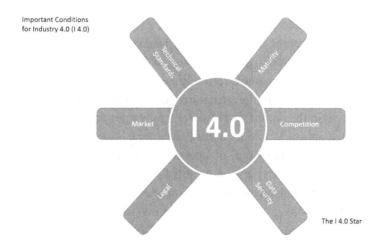

mean software. The software share in products is changing the markets and products faster in a rapid growth scenario. How can the company recognise and react to the changes without a doubt? This question sets a new focus on agile approaches. In software development, for example, Scrum and Kanban software are known as agile development methods. Kanban is widely used in industrial production. Now, these methods are moving into all organizational units and structures of a company, even in areas beyond production. This makes it clear that the successful introduction of lean management as an interdisciplinary company philosophy is the prerequisite for Industry 4.0.

But is now every industrial company becoming a software producer? Or should the relevant development area be outsourced to a subsidiary or sister company? The possibilities offered by current technologies seem to be boundless. Countless questions arise here. Which path would represent the most successful strategy with the best future potential? The evaluation of the various strategies according to their meaning or uselessness must always be questioned centrally. True again: The good development of customer benefits and business cases is a lasting improvement to the success opportunities for a transformation to Industry 4.0.

Company Organization of the Future: Where Is the Journey Taking Us?

In companies, structures have developed over decades. Hierarchies and responsibilities are defined. Successes lead to the expansion of organizational units. New employees are hired and the structures continue to grow. Rules and processes are adapted accordingly, but business areas are very rarely changed. "Never change a running system or a winning team" comes to mind. A principle we all like to follow. Therefore, changes often concentrate on a particular aspect or business area. This makes efficiency control and cause analysis easier: If expectations are not met, the causes can be quickly determined and sensible, necessary adaptations can be implemented.

In Industry 4.0, however, the holistic view is required, as mentioned previously. Does one have to say goodbye to the idea of implementing change in small, manageable steps? Systems must certainly be developed in a multi-disciplinary manner in the I 4.0 era. Therefore, structures have to be changed simultaneously at several points within the company. However, companies are often organized in the form of specialist areas. A "silo mentality" developed over decades, which is now an obstacle in the implementation of interdisciplinary approaches. Rethinking is important now. It is often advisable to establish a program that controls the various projects in the individual areas. The program should ensure that the necessary changes from the various projects are implemented in a coordinated manner. With regard to the question at the beginning, changes can be implemented iteratively in small steps if, in return, a road map is established, describing the individual implementation phases. The said road map could be maintained within the program as to not lose the focus on the overall goal. It would also be sensible to deal, among other things, with the requirements and scoping management in this institution in order to prevent moving target effects professionally. The question of which cross-sectional functions are established here will certainly have a critical impact on the entire transformation program. Other boards on the topics of quality, technology, organization, culture, etc., are possible.

As this is about the development of systems, systems engineering will gain in importance. This discipline can certainly already to be found in the corporate environment. In the future, systems engineering will be widely used in all company sizes.

This raises the question of the optimal corporate organization of the future. Does a functionally organized company still make sense or rather a matrix organization? Or rather agile management as a further development of lean management in order to find appropriate, up-to-date answers to globalization? To be honest, there is no definitive answer. It is, however, clear that companies must also reinvent themselves. For many experts today assume that our current understanding of products and services will no longer fit within the framework of future-oriented, intelligent production. This also includes the relatively strict division of the fields product development, laboratory/testing, planning, production, and services.

Flexible and Equally Risk-Based Process Management

Over the past decades, the German industries have always been able to understand the conception, design, and production of modern, high-performance products based on business processes. They have almost exhausted the standardization potential within the scope of the respective possibilities. As a result, there have been or will be competitive products, which are always at the top of the market with regard to their product costs compared to the world market, but overall they have an extraordinarily good price/benefit ratio. This is often the reason why many German companies are market leaders. Embedded software components contribute to the success of the company with growing importance in order to successfully manage the increasing product complexity. If, however, the software share of a product is continually increasing, the software mutates from provision to the central product component: a secondary issue becomes the main component. Industry 4.0 describes the relevant transition with a focus on this effect. In order for the software to play its true possibilities, however, it requires networking. This effect was already seen several decades ago within the framework of office communication (from the standalone PC in the office to a PC connected to the company network and finally over the internet to the rest of the world). In relation to products, however, this requires a completely new consideration. Networking, so far, has not been a part of current product developments. How do the development and production processes change when, as mentioned above, they were conceived, introduced and standardized with so much accuracy? Networking over the entire life cycle seems to be important. Flexibility and adaptability will gain in importance in the case of process questions in Industry 4.0 in order to meet the constant changes within the framework of globalization. Standardization is good. However, it can also be restrictive and rigid, which can quickly drive successful companies out of the market.

Procedures, Tools and Applications Need to Reinvent Themselves

New technology and techniques were developed by the technology experts. Some, like the embedded software, I already mentioned. The resulting continuous increase in tool performance led to model-based work. The term "virtual reality" is mentioned here as an example, which makes it possible to experience planned, complex systems in a way that was not possible before. 3D printers are also mentioned in this context. This new way of working demands even more intelligence and networking in the future. Thus, collaborative working will be the way to go. In relation to models, however, this means that these are understood equally by all experts of all knowledge domains. Today, models exist only for the respective engineering discipline. It is like a language that is understood and spoken only by these specialists. Do you know this effect when you join a circle of participants with a special topic you do not know about and you do not really understand what they are talking about? The same is also true for data. If methods have been developed with focus on the company divisions, it is obvious that the relevant data is created and managed optimally for the said area or process. Often, this data can only be used in parts, if at all, for other business areas. There is simply no understanding to correctly interpret the data. In the future, models — or rather a model — are needed based on a discipline-neutral norm, which can be understood by all experts regardless of the qualification background. Today's existing specific-purpose processes need to be expanded and harmonized fundamentally.

About the Inflation of Complexity

Bjarne Stroustrup once said, "I've always wanted my computer to be as easy to use as my phone; my wish was fulfilled: I no longer can use my telephone" (Stroustrup, n.d.) No matter what technology product you look at, they all have in common that their complexity has increased enormously. This explosive growth of complexity with inflationary dimensions is, in my opinion, the greatest challenge; today as in the near future. This is accompanied by a growing complexity of processes in all industrial areas, from planning through development to production. The resulting tension from preferably stable processes in design, manufacturing and elsewhere, on the one hand, combined with a great and versatile product variety with intelligent functions on the other hand has so far been cleverly solved using embedded software components. Companies that are most adept in this discipline are generally also leaders in their market segment today. This lead will not last forever. However, the experience from integrating the embedded world into products could be of benefit to conquer upcoming tasks. In concrete terms, this means that business models, procedures, processes, and skills must be aligned to networked, intelligent systems. To show or offer a useful additional benefit for customers, which would certainly lead to a further integration of software into all product types. Perhaps this is the solution to get a grip on the complexity and equally to implement intelligent solutions in which ultimately the information is the central aspect. Perhaps it would then also be possible to run even the most complex processes so simply that it would not require any human interaction at. Products would then not only be exceptional and of high-quality, but simply brilliant. You could even take that literally.

Rapid Change in Professional Profiles

Many employees in industrial companies link the transition to the I 4.0 era with comprehensive changes in the workplace. The central focus here is on the new requirements for the employees themselves. For example, through the various stages of evolution in the industry, the metal worker became an industrial mechanic, and ultimately a mechatronics engineer in Industry 3.0. The job descriptions for mechatronics engineers is also changing at a rapid pace. The new professional title has not yet been found. This effect is valid in all areas.

In the future, new employees will not be able to be deployed easily or quickly into the production processes. Although there are the requirements for Industry 4.0 that the future systems should feature a high usability and not require long introduction times. However, in my opinion, the conclusion that high usability guarantees short or no training times at all is drawn to quickly. It can be seen in practice that modern systems, which are controlled via touchpad or by voice, often have to be adapted to the person concerned and they have to be customized. It must also be taken into account that the new employee must also be involved in the then much more complex processes due to in-depth integration with other systems. They must understand the interaction between the systems so that they can correctly evaluate any consequences based on their input before issuing the commands.

What do we know about future employee requirements? To be honest, not all requirements are known yet. It is, however, certain that IT knowledge and skills are increasing as part of system programming and configuration. It is very likely that this discipline will enter the industry in every profession, assuming this has not already happened. This, in turn, affects the professional image of computer scientists. They will have to deal even more with the terms complexity and information. Their role will change from the former programmer to that of an information manager. The trend is already clearly visible today.

Data Security Is Often Underestimated

When people are discussing Industry 4.0 they touch upon many aspects. Often, however, the importance of safety is neglected. I often hear the comment that IT security is very well a subject that is being discussed. This may certainly be true. My statement, however, considers the appropriate measures for a comprehensive security policy. Since all systems are being linked and communication can take place via different locations or with partners and customers with regard to relevant production data, new points of attack arise for the respective organization. Potential security risks include:

- Spying on production or customer data to gain competitive advantages for competitors.
- Import of malicious software with the aim of paralyzing the relevant infrastructure for an extended period of time.
- Slight change of production data to generate defective production. The aim here could be to reduce customer trust in the company concerned.
- Change of environmental information, such as humidity or temperature, so that the production plant is damaged due to insufficient adaptation or the reject rate increases.
- Targeted bombing with manipulated information focusing on a particular industry to eliminate it.

These and other new threats must be included in regular security audits. Various solutions can be envisaged in order to meet these new threats. These range from the complete foreclosure of critical business areas (IT), with dedicated access to proven gates, which are under special surveillance, to the deployment of comprehensive, active security solutions (beyond the ordinary virus scanner) and intelligent monitoring systems for the increasingly complex IT infrastructure.

In addition, there is still the aspect of data availability and completeness within IT security. In the event of a system failure, the information must be available quickly. In itself, this topic is really not new. However, we may need to think about quite different amounts of data and time periods than is the case today. There could be networked systems, for example, which would come to a standstill after a few minutes without a new information supply since greater inter-buffering of data would not be possible in this scenario. This case could occur, for example, if the system would make decisions other than foreseen by real-time status changes, initiated by other systems connected to it.

EXAMPLES OF SUCCESSFUL IMPLEMENTATIONS OF INTELLIGENT PRODUCTION SOLUTIONS THROUGH DIGITAL TRANSFORMATION

Industry 4.0 sounds interesting, but can the approaches be transferred to practice at a reasonable cost? A question that I always hear in projects and in my daily work. My answer to this is: Yes, they can! However, one must approach the topic with the necessary experience. Thus, I would like to present some successful implementation examples on the topic of digital transformation within production and logistics.

Example 1: Lot Size 1

Lot size 1 is one of the stimulus themes in the context of digital transformation. In this case, it often prevails that lot size 1 is an ideal type goal, which is not really realizable. The resulting high costs within the establishment would ultimately hinder such projects.

Already in the first part of the series I emphasized the importance of digital planning and production processes in the I 4.0 era. And these processes need an optimal interlocking into the real world of production and engineering. The following points are particularly critical for success:

1. A perfect combination of software systems for product design and production automation.
2. Optimum coordination of the operating, machine and process chains without interrupting interfaces or with a high degree of harmonization.
3. This allows the third criterion to be implemented: smooth communication between all systems and components along the logistical value chain within a company or group of companies.

For me, the above points are the three key core objectives of a digital transformation in a company. On this basis, savings potentials can be realized, which can be as much as 30%. In doing so, often all parties involved along the value chain will benefit; the company, the business partners, the customers, the employees and the users.

I would like to present an example from medical technology. In the transformation of classical processes into the digital world, one did not choose the best way to develop digital processes with the available data, but developed a completely new approach without considering the established systems, paths, rules, data, views, etc.

Specifically, it was about developing a method for the production of knee joints, which should offer cost advantages as well as advantages for the patient along the healing process compared to classical production methods. In summary, the method development can be condensed into the following three phases:

- Problem analysis with exact formulation of the requirement as an analysis result.
- Derivation and construction of the necessary data framework.
- Conception, development, and establishment of the processes together with the supporting systems.

The goal was to produce individual knee joints in series instead of, as usual, in a certain grid size (small, medium, large). The task of the physician was then to bring together the dimensions of the affected joint with the circumstances in the patient. As a rule, this led to a high degree of adaptation and complicated healing processes as a consequence of these necessary measures.

The advantages in the production of completely individualized knee joints are obvious for the patient compared to the previously explained method: necessary adjustments by the physician to the bone structure are clearly reduced. The associated healing process is accelerated considerably.

Individualized knee joints have been produced ever since artificial knee joints have been made by specialists in factories. Within the framework of the medium-sized enterprise, the innovation consists in manufacturing the products industrially in serial production with lot size 1. In the end, artificial knee joints are obtained which are tailor-made and matched to the patients concerned, in a fraction of the production time of a manufactory.

How does this production process work? The bone of the patient is first measured so that a three-dimensional data model of the relevant joint is present at the end of the measuring process. From this, a prosthesis model is built. This means that the data and the production process can be checked for correctness on the basis of the test results since every individual production process delivers individualized products. This means that a sample test would not be able to meet the quality requirement. When the results fully satisfy the requirements, the actual production is started by importing the 3D data into the CNC milling machine. The product is available after a few hours.

The process described here is implemented by networking various IT and application systems from the fields of 3D modeling, PLM, production control and production automation.

Example 2: I 4.0 Drawn by the Eye

For Industry 4.0, the holistic approach, i.e. the horizontal and vertical integration, is the recommended path along the digital transformation. Does this ultimately reinvent every process and organizational structure completely? Often this seems to be the necessity, but this rule is not compelling, as my second example shows.

The second example is about a leading supplier of high-quality sinks and kitchen fittings. The range also offers waste disposal systems and accessories around the sink. The company group employs a total of 1,400 people worldwide, of whom 1,040 are located in Germany.

For the group's European logistics center, the following strategic priorities are of particular relevance for daily business transactions:

1. Ensuring a high level of delivery availability, which can meet current requirements at short notice.
2. Establishing process excellence with the goal of steadily improving responsibility and quality awareness as well as customer awareness.

For the group, quality is ultimately the all-important factor, to which all other objectives are subordinated. This aspect does not represent part of a clever marketing strategy, but the lived realization that without generating outstanding product and service qualities no sustainable business success can be generated. It is therefore not surprising that over 600 suggestions for improvements are communicated annually by the individual teams within the framework of their CIP initiative (continuous improvement process) and are processed by the quality management department. In this way, the processes and the work environment are continually being optimized and adapted to the current situation.

A central aspect is the optimal fulfillment of all customer requirements by means of an IT that is deeply and optimally integrated into the business processes. The IT landscape is very SAP-oriented. Almost all systems from the SAP portfolio, such as CO, FI, SD, MM, SRM, QM, PM, etc., to name a few, reside in the data center next to the company headquarters. The globally distributed locations are connected to the data center. This means that no server applications are running on the sites beyond the local IT infrastructure. This results in relatively high requirements for the connection of all branch offices, which is implemented without exception via fixed lines. But today, due to technological developments, online access is more cost-effective than it was the case decades ago. Furthermore, this type of access offers advantages over matching, replication and buffering methods of local data silos, since the latter are always at the expense of data actuality and quality.

The company does not shy away from comprehensive and far-reaching adjustments to the SAP application landscape, which go far beyond trivial ABAP code changes. On this basis, the IT team, together with the logistics department, implemented a Pick by Voice solution with the aim of optimizing round trips and/or downtime reduction, employee input, and utilization of workstations along the consignment processes. In the background, these SAP application systems work on the consignment orders that are currently being processed by the logistics center. Depending on the structure of the order, the disposition differentiates between three main types of planning, which differ from one another in the degree of planning detailing. The order data is then transferred from the data center to the logistics center and ultimately to the relevant employee on his headset. A voice communicates the next work steps to the worker, which are acknowledged directly afterward by them using the headset. These success reports are transmitted from the logistics center to the data center, whereby the SAP applications update the order data and simultaneously transmit the data for the next job. The dialogs, the communication, and the data comparison always take place online. No essential data is buffered.

However, the extensive technology integration should not lead the reader into the fact that the principle of innovation would be handled at any price in the company group. Rather, decisions are drawn by the eye and made with a healthy common sense whether the use of new technologies actually opens up optimization potential. Thus, within the described consignment process, simple color coding is used to control processes: Within SAP planning, orders that belong to a consignment group are marked in color. The Pick by Voice System informs the employee about the coding by means of the operating instructions, which is converted by the user by attaching a magnetic sign to the wagon in the appropriate color. In this way, the wagon group can be easily identified and treated accordingly within the logistics processes.

Example 3: Total Integration

In I 4.0, product design, production planning, and production engineering are particularly important along the value chain, as can be seen in the first two examples. Compared to classical production methods, costs are initially higher in relation to the total cost of production. This fact makes many companies hesitate to drive the digital transformation for themselves. If, however, optimum production is carried out during the first phases of the production process, the following advantages can be achieved in the subsequent on-going production compared to classical approaches.

In this context, a highly integrated automation, which points to further optimization potentials within the production automation, is particularly interesting. There are different terms used in the market for this approach. Essentially, it is a question of establishing an automation, not just within a hierarchical level, but across at all levels (company level, management level, control level and field level) within a company (we recall horizontal and vertical integration from I 4.0). So the creation of total integration. Two strategies are pursued:

1. Further increase in performance and reactivity through production control systems (or also called Manufacturing Execution Systems/MES).
2. Improved production transparency through further and more efficient networking so that the top management can directly access the production line and see data in real-time and make changes.

For this purpose, an implementation example from the field of glass production will show what flexible and intelligent production can look like. In the production plant, a high degree of automation and optimum energy efficiency were already ensured during the erection of the production facility. Both aspects offer the supplier of glass products decisive competitive advantages with regard to the resulting cost structures on the market. As a result, highly integrated automation solutions and powerful systems for heat recovery were installed in combination with comprehensive energy management as part of the design and construction of the production line. The glass production and all subsequent processes, starting with the delivery of the raw materials through to the melting process to delivery, are completely automated. In order for the system to be able to produce without interruption throughout the year, all systems in automation, energy management, installations and drives are networked with one another. Numerous sensors along the production line deliver all relevant data in real-time, so that in the event of deviations, immediate corrections can be initiated, which are immediately implemented without time loss by the respective systems. In this way, the cost advantages assessed during the planning phase can be realized during operation.

Example 4: Heavy Weight Arguments

The networking of the virtual and the real working world is becoming increasingly important. The heavier the products to be manufactured are, the more urgent the implementation of this requirement seems to be. This example involves the production of an 85-ton precision optical machine, which processes telescope mirrors with a diameter of up to 2 meters for space research. The grinding processes carried out by the machine must have an accuracy of 30 nanometers. For this purpose, especially integrated measuring procedures are provided, which ensure the required tight tolerances at the production end of a mirror.

Such an optical machine with the aforementioned dimensions, requirements, and qualities can no longer be successfully produced without prior comprehensive planning and simulation. First, the machine was designed in a PLM system using CAD applications. This results in a virtual machine, which is the basis for the necessary simulations in the subsequent phase. The findings of the simulations lead to further optimization in the 3D model. The iterations between CAD design and simulation are repeated until the simulation result meets the requirements agreed with the customer. Only then the machine is actually produced, individually.

But also the finished product itself has I 4.0 competence. This is because high-performance hardware and CAD software are also used here, in combination with integrated sensors and complex drive components, to sharpen mirrors with the required, low tolerance values in a significantly reduced throughput time. This new spacecraft telescope allows astronauts to look into space up to 13.5 billion light years, even faster than before.

CONCLUSION

Four examples show four different approaches and also solutions. And how such a digital transformation may look like. All four solutions have one thing in common: success. This should be encouraging. Companies that are left behind may be pushed out of the market. These four examples also show that there is no patent recipe in this discipline. So how do you find good solutions? In the context of I 4.0 implementation projects, the holistic approach is certainly a central point. Because every detail in the organization or in the processes can be important or critical to success in terms of feasibility and possible risks. The benefits should always be questioned for all the changes and improvement approaches envisaged. However, the potential benefit is not limited to time and cost reduction. Rather, detailed information for the customer, better services or intelligent products optimally adapted to the application environment should also be taken into account as possible potentials in a valuation.

REFERENCES

Allianz. (n.d.). KIT Industry 4.0 Collaboration Lab. *Allianz Industrie 4.0.* Retrieved from: https://www.i40-bw.de/de/100orte/imi-am-kit/

Carnegie Mellon University. (n.d.). Retrieved from: https://www.cmu.edu

ETH Zurich. (n.d.). *ETH Zurich.* Retrieved from: https://www.ethz.ch

KIT. (n.d.). *Industry 4.0 Collaboration Lab.* Karlsruher Institut for Technologie. Retrieved from: https://www.imi.kit.edu/2754.php

McKinsey. (n.d.). *McKinsey & Company.* Retrieved from: https://www.mckinsey.com/

Stroustrup. (n.d.). Bjarne Stroustrup's FAQ. *Stroustrup.* Retrieved from: http://www.stroustrup.com/bs_faq.html#really-say-that

VDMA. (n.d.). About the VDMA. *VDMA.* Retrieved from: http://www.vdma.org/

APPENDIX

The Future of Digital Transformation

Our world is constantly changing. It has become volatile, uncertain, complex and ambiguous. The trend towards comprehensive digitization and the convergence of the IT, industrial and financial markets with regard to products, processes, and services lead to the omnipresence and unrestricted usability of the internet. From the original means of communication for people via computers, an "Internet of Things" (IOT) has developed, which helps people in their activities imperceptibly and leads to a fundamental social change. This is expressed first and foremost in a natural human/computer/machine communication style, which is based on understanding and dialog and requires the interconnection of several communication channels. The so-called digital transformation under the term "fourth industrial revolution" will lead to a mutual fertilization of inventions and innovations driven by the elimination of the location principle in the market, digital business models, modular value chains and new competitors. This change is still at an early stage, but it is already visible today in a wide range of human spheres, societies, technology, economics and value systems.

The individual could be a powerful driver, even a "mega trend" in the 4th Industrial Revolution. Every industrial revolution has profoundly changed the world. After total computerization and the creation of a world economy as a result of the third industrial revolution, the fourth industrial revolution is expected to lead to a globally networked community of values. Since companies have begun to operate globally, they must adapt to the conditions of the various regional markets and cultures of the world in terms of products and services. At the same time, networking between humans as individuals, also through the IoT, plays the decisive role. Instead of using computers, the IoT now helps people in their activities imperceptibly, while the individual is devoted to their subjective and emotionally emphasized perception of products, services, and life, education, or professional activities in general. This results from the declining dependency of the individual on traditional ties and norms, reinforced by the general prosperity growth, and is accompanied by a corresponding change in ownership and user motivation, emotion, experience, and enthusiasm are the main focus. More and more people are engaged, sharing images and content, commenting on social networks, expressing recommendations and comparing certain brands, products, and services, i.e. looking at them as "friends".

In addition to the positive effects of digitization, there are also major risks and dangers. The McKinsey Global Institute, for example, estimates that by 2025,

Figure 3.

Market Changes

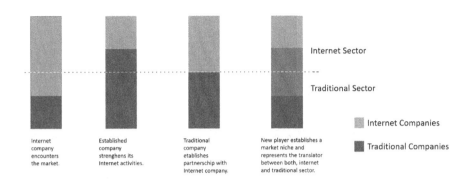

Internet Sector

Traditional Sector

Internet Companies

Traditional Companies

Internet company encounters the market.

Established company strenghens its Internet activities.

Traditional company etablishes partnerschip with Internet company.

New player establishes a market niche and represents the translator between both, internet and traditional sector.

Research shows four possible scenarios in terms of market changes. For example, the triumph of the Internet elicits all of these presented scenarios in the chart above. Similar scenarios are now also possible concerning the advent of Industry 4.0.

robots and machines could replace an estimated 140 million knowledge workers with intelligent technology (McKinsey, n.d.). Already in the 1970s, one believed in the power of artificial intelligence and the fact that there will be intelligent robots, paperless communication and humanless factories over the next twenty or thirty years. We can see today how far we are from it. The main problem is that people have not yet taught robots and machines to "think". Humans can cope with inconsistencies, differences, or ambiguous information that seem difficult to understand or even unacceptable, but machines cannot do so yet. The future will tell whether machines will take control of us humans or not. But what is already clear today is that they are increasingly pushing people away from the routine labor market. What initially appears to be menacing, on the other hand, is the chance for changes in professional images and the resulting change in perspectives.

In the age of Industry 4.0, the dynamic networking and the independent (autonomous) communication of the individual components of a system over the internet gain increasing importance. A "Cyber-Physical System" (CPS) refers to the combination of informatics, software components with mechanical and electronic parts. These communicate via a common data infrastructure, such as the internet. The CPS are characterized by a high degree of complexity. This approach is directed only at the "object" (machine or computer) as the center of the consideration. However, due to the rapid development of industrial networking technology, as well as social networks and online services, the role of "Human Machine Human" (HMH) and "Human Computer Human" (HCH) communication is growing. The main focus of the examination passes from the "object" (machine, computer) to the

Figure 4.

Development of Customer Benefit

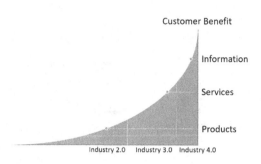

Maximum customer benefits can be realized with the in time delivery of the correct and adequate information.

"subject" (human). Recent trends point to the fact that the border between "being online and being offline" is subsiding for people. In this way, people's perception of reality in space and time changes. Material and immaterial worlds merge. Real-time applications, supported by realistic visualization technologies, make it possible for people to make invisible phenomena visible and validatable early on, thereby realizing new product features and functions.

In the future, companies will need the overall consideration of the company's development from the outset. Not only the technical feasibility, in particular, the country-specific, economic and social boundary conditions have to be taken into account. In doing so, goals are to be set in a company-specific manner and in a timely manner in order to adapt to the increasingly rapid successive innovation cycles. The knowledge and technology generations are now changing every three to five years. Furthermore, a traverse implementation by decision makers "top-down" and employees "bottom-up" is important. Changes are lived in the "breeding grounds" of a company and the measured value is the added value. The creation of a human-centered organization means to realize brain-oriented work processes in dynamic value-added networks. These are characterized by rapid analysis and optimization cycles. The digitization is no longer "out-sourced". It will simply be part of our existence, in the daily business as well as in everyday life.

3D Printing, Artificial Intelligence, Virtual Reality and Networking (until 2023)

When we first look at the next five years (2018-2023), the next innovations are clearly visible for many market participants. One of these is 3D printing. So far this still represents a niche. This is mainly due to the raw materials to be processed. These are currently not suitable for all possible applications. Consider, for example, requirements such as stability, robustness against wear, acid resistance, temperature or medical compatibility. Currently, there are limits to the substances used in 3D printing. However, these limitations are eliminated step by step. In recent years, the number of materials used has clearly increased. This trend is likely to continue in the near future. The materials based on polyethylene or polyester are combined with other materials in order to improve the material properties in terms of processing and subsequent use of the product produced. For this reason, many people predict a golden future for this technology.

Just imagine you were the captain of a cargo aircraft with a cargo company. You have just landed in North Africa, for example, at the airport of Cairo. The goods destined for the region have been discharged. After that, it is off to New Delhi. But at the cross-check before the start, your crew noticed that a component was damaged, probably during the landing in Cairo. Without a successful repair or replacement of the worn component, the tower will not release you for takeoff. The spare part is not available in Cairo. This would make it impossible for you to perform the next flight order. However, with the help of 3D printing, the situation looks quite different: the manufacturer of the component sends the design data to a certified service partner on site. They print the component based on the design information. A few hours later, the component has been replaced and successfully tested. Nothing is holding you back from starting your next leg. You might land in New Delhi with a delay. However, there are good chances to successfully complete the transport flight. Today, a situation like this would probably require a replacement for the defective aircraft in Cairo. 3D printing is a technology that will dramatically change the production of goods. That is why it is often assigned to the disruptive technologies and its business processes to the disruptive business processes. They are likely to displace classical technologies and their business processes. Does the future belong to 3D printing? Many market players are confident that the 3D printing technology has fantastic prospects and holds a lot of potentials. In my article, I will come back to this question.

In addition to 3D printing, there are many other technologies that will make further progress in the coming years. I see a lot of potential here, for example, in sensor technology, networking, artificial intelligence or virtual reality. The glasses that you have to wear to immerse yourself in virtual environments are already well-known. A new technology, however, is special contact lenses, which makes glasses unnecessary. In the contact lenses, special chips are incorporated, which project image data directly onto the retina. The data is sent by radio from a computer. For the viewer, the virtual is mixed with the real image. The danger is that they can no longer distinguish between what is real and what is virtual. Only by removing the lenses would this be possible without a doubt. However, this solution offers some significant advantages over the VR glasses. Actually, many solutions are already available. But they often act as an island solution. However, the actual potential will be shown once these singular solutions are highly integrated and networked.

Within the framework of the High-Tech Strategy Industry 4.0, many developments have been and are being advanced. However, there are some areas that have been neglected conspicuously. These include topics such as usability, product lifecycle, recyclability and the materials themselves.

Intelligent Materials are the Key to the Actual I 4.0 Potential (2020-2030)

I would like to focus on the materials. Is it not surprising that there are comparatively few new technologies in this area? For me, it seems as if the I 4.0 developments were realized around the central theme. However, it should be clear that the further development of materials technologies is the real key to digital transformation. Here is a small motivation. What if we could control the construction of an atom? In my diagram, you can see a hydrogen atom. By means of an apparatus, a controller, one would be able to produce a positive and a negative potential, which ultimately would produce an electron and a positron. The electron migrates to the second electron pathway and the positron into the nucleus. A helium atom has been formed. Does this sound too fantastic? In fact, this process cannot be implemented with today's technologies. In attempting to add another proton to the nucleus, the protons present in the nucleus would repel this nucleus. But the thought is really nice: Perhaps, finally, the feat of transforming lead into gold would be possible?

We remember that researchers at the Leipzig Max Planck Institute were able to prove that DNA can be stored for several hundred thousand years by sequencing a 300,000-year-old bear DNA (ETH Zurich, n.d.). This gave scientists at the ETH Zurich, Switzerland, the idea of storing data in a DNA. They succeeded in 2015. This makes it possible to manage data on a molecular level in principle. And one can control molecular structures.

The current research at Carnegie Mellon University, USA, is based on this: Under the term "programmable matter", DNA strands are artificially linked and their properties are changed (Carnegie Mellon University, n.d.). This has already been achieved in the laboratory. Here, any number of DNA strands can be connected

Figure 5.

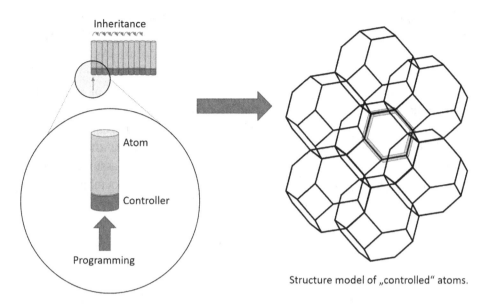

Structure model of „controlled" atoms.

and formed. In this way, tiny components could be developed on a nanoscale, such as, for example, microscopic small gears (see image). After this success, the real goal is to master the technology of programmable atoms, which are often called catoms or conatoms.

Similar to the initially described motivation, each atom gets a controller. The atom can be controlled using the respective controller. That is, the properties of the atom can be changed by man. This has already been achieved in the laboratory with one atom. In the coming years, attempts will be made to form a series of catoms. In doing so, the first atom is programmed, with its neighbor catoms taking over the information in turn. This process is reminiscent of the inheritance of properties from object-oriented programming. Thus, it would already be possible to control a material with regard to its nature, its appearance, and its properties. However, the arrangement in question would be too inefficient. The aim is therefore to form a hexagonal grid of catoms. This would allow you to program a column from a cube without using the usual tools (see image). Or how about a valve, an engine, a hydraulic system or a plane? The transformation would take some time because of the inheritance since it requires several iterations for the metamorphosis. In principle, this method could be used to generate any object. And in any place. Today, products often have to be created at the dedicated production sites. Similar to 3D printing, it would also be conceivable here that this rule will no longer apply in the future. The need for special tools, special production facilities, and infrastructure for product realization will subside. Products could be created at the place of use. It would simply require transmitting the necessary software and the product-related data. And when the product is no longer needed, it is reprogrammed. The recycling topic would therefore also be under control with this procedure. As interesting as material properties are in production, the application possibilities in everyday life would also be revolutionizing. The materials would have a kind of intelligence, which would memorize their programmed properties. So it would be possible to form the material with bare hands. If one left it to itself, it would change back to its original form. Imagine: finally, no more dents on the beloved car after a bumping into something when parking. These intelligent materials could assume properties which today are hardly or not at all possible to be realized in combination, thus significantly expanding the range of application of products:

- Particularly light and thin but at the same time extremely stable and flexible in certain situations.
- Depending on the situation, permeable, semipermeable or absolutely dense.
- As well as the situation-dependent adaptation of the product form.

I reckon that in the beginning of the middle of the 2020s detailed knowledge and stable procedures from science and laboratories on the subject of catoms are available to the extent that this innovative technology will find its way into the modern factory.

Can you remember my question regarding the evaluation of 3D printing technology? Perhaps you were also of the opinion that 3D printing holds a golden future. However, with regard to the programmable atoms, the technology might face difficult times from 2022 onwards. One disruptive technology is eaten by another, comes to my mind. This allows us to add another feature to the already known properties: disruptive business processes and technologies have cannibal inclinations.

Challenges for the Industry

The digital transformation provides high challenges for the industry and the manufacturing sector at various levels. Here the question arises for management how to approach this highly complex topic in a meaningful way. It is relatively unlikely that companies will be able to successfully evade the resulting issues. In order to support the economy, the IMI (Institute for Information Management in Engineering / Institute of Engineering), an institute at the KIT (Karlsruhe Institute of Technology / Competence Cluster in Technology Topics), led by Dr. Dr.-Ing., Dr. h.c. Jivka Ovtcharova launched the "Industry 4.0 Collaboration Lab".

On February 19, 2016, the "Industry 4.0 Collaboration Lab" was awarded by the Allianz Industrie 4.0 (Alliance of Industry 4.0) Baden-Wuerttemberg as one of the

"100 Places for Industry 4.0 in Baden-Wuerttemberg" (KIT, n.d.; Allianz, n.d.). The solutions at the IMI Institute are above all a test and qualification laboratory. In this laboratory, realistic scenarios can be carried out using the one's won company data sets and handy solutions for one's own business. The target group is technicians and engineers in the production area, but also board members, managing directors and production managers, who seek to get an impression of smart production.

In line with the credo "SME Meets Research", the lab offers companies the opportunity to test their ideas and products at an early stage and to become familiar with the work in 3D environments. An appropriate training of employees is also possible here.

The "Industry 4.0 Collaboration Lab" systematically includes other data sources as well as development data. The laboratory provides interlocking software and hardware structures. This allows individual requirements to be better considered and development times shortened. Through the use of virtual engineering, which supports the entire product lifecycle, from a process point of view as well as IT system view, new engineering methods are being developed. These systems also serve to reconcile, evaluate and secure the results. Modern information and communication technologies such as CAD/CAE, PDM/PLM, web, cloud, and virtual reality are used.

In the context of the "digitalization as a daily business", opportunities, prerequisites, obstacles and potentials can be worked out by means of a so-called virtual image for Industry 4.0. The virtual image in production is a solution approach for the optimal operation of a factory. This includes the 3D geometry of the physical environment in conjunction with the factory and process properties. The interfaces to external systems and simulations with consideration of all relevant resources and processes are decisive. The interfaces can be based on the "Internet of Things" concept. Operational concepts for production plants can be validated in real-time: manual and automatic operation and configuration via intuitive man-machine interfaces (such as web interface, haptic interaction devices, etc.), for example. This makes it possible to make decisions on the basis of real-time information in the complex factories. Simulations could also be used to generate future forecasts based on real-time data from production. By merging real and virtual environments, an approach to the vision of automated, intelligent and virtual commissioning of a complete production is possible.

The KIT/IMI Industry 4.0 Collaboration Lab can particularly support the industry in the creation of new core competencies that will be needed in the future to deal with an increasingly digitized, automated, networked and complex world and to find a valuable place in it. Probably the most important fields of competence in the digital transformation are as follows:

1. Real-time data acquisition, processing, and evaluation of different, heterogeneous and unstructured data sources (also in the sense of Big Data).
2. Intelligent mapping (algorithmization) of decision-making and action sequences as the basis for new business models and processes.
3. The ubiquitous education and qualification of all participants, in order to integrate the digitization into daily business.

Chapter 2
Conceptual Foundations of Creating Sustainable Development Strategy of Smart Cities:
Environmental Aspect

Olga Burmatova
Institute of Economics and Industrial Engineering of the Siberian Branch of the RAS, Russia

ABSTRACT

This chapter is devoted to the study of the role of ecological subsystem in the structure of the sustainable development program of smart city. The author suggests the logic of building the environmental strategy of the city as a long-term landmark of its sustainable development including the environmental mission, vision of the future, goals and priorities, programs and their implementation, target indicators for assessing results, and consequences of realization programs. Certain attention is paid to the city as an object of research with a focus on environmental problems. The characteristics of the factors affecting the development of the ecological situation in the city are shown. A system of criteria and indicators that can be used to assess the impact of the planned environmental activities is proposed.

DOI: 10.4018/978-1-5225-3996-4.ch002

Copyright © 2018, IGI Global. Copying or distributing in print or electronic forms without written permission of IGI Global is prohibited.

INTRODUCTION

Smart City Concept combines various factors of development of the city into one system, including economics, management, energy, transport, environment and population. At the same time it relies on the increased role of human capital and the strengthening of the importance of information technology in the urban environment. The obligatory feature of any Smart City is its long-term sustainable development that can be achieved, if the balance of the three spheres - economic, social and environmental – is enforced complied. That means that any administrative decision shall be without prejudice to any of the components of the city subsystems - society, the economy, environmental conditions, etc. Therefore, for a Smart City is extremely important to have effective management and analytical tools to promptly predict potential negative externalities, as much as possible to internalize them and manage them. One of these tools is able to serve, in particular, strategic planning and management. More and more cities are on the path of development strategy, which is based on the concept of a better future and available opportunities and threats.

Obviously, the easiest way to create a Smart City from the ground up on a single project. It is much more difficult to transform an existing city, especially large one, in the Smart City. Most often within existing cities projects for creating "smart" blocks are being developed and being implemented, and these or other elements of the "smart" cities are gradually introduced. In world practice of the functioning and development of cities such attempts are quite numerous (mart Cities, 2014; Hollands, 2008; Songdo, n.d.; In Japan, 2014). For example, Vienna, Barcelona and Copenhagen are included to the category of Smart Cities.

Usually in existing cities the idea of "smart" cities are being introduced gradually and consistently, from "smart" solutions to improve the urban environment and the mobility of the population with further coverage of the various spheres of the city, including economics, management and ecology, to create ultimately in the long term Smart Model of life in the city.

Attempts to create a Smart Cities are undertaken in Russia as well (The Smart Cities, 2014), in particular, in Novosibirsk. For example, under the Program re-industrialization of the Novosibirsk Region's economy until 2025 (Program, 2009) the concept of Smart Region is proposed. It is aimed primarily at improving the quality of the urban environment and mobility through the use of information and communication technologies. The concept is aimed at improving the urban environment, and assumes management of the city and its economy, social sphere, transport system, the environment and life through smart technology (Program, 2009, p. 44-48). The aim of the concept is to make the people living in the city as comfortable and safe. One of the important elements of the Smart Cities is ecological.

The author focuses attention on the integration of environmental aspects into the strategy of the city (with an emphasis on the big city).

The aim of this study is to develop a methodological and methodical fundamentals of formation of environmental strategies within the framework of ensuring sustainable development of Smart City.

Realization of this goal involves the following major tasks:

- Consider the city as an object of research;
- Show the place of ecological subsystems in the structure of socio-economic development of the city and to analyze its relationship with other urban areas;
- Show the possible directions of predicting the environmental impacts of economic activity in the city;
- Formulate the most important directions of formation of city environmental strategies and analyze the main factors of its formation, including spatial, industrial and institutional aspects of the selection variant conservation strategy;
- Propose a system of criteria and indicators that can be used for assessing the impact of the planned environmental activities;
- Identify conditions for the implementation of environmental policies;
- Carry out the practical application of the proposed approach in the development of socio-economic development of a particular city (on the example of novosibirsk).

The City as an Object of Research

The range of problems covering various aspects of interactions in the system "big city – the environment", is very broad and includes, in particular, such problems as the formation of the favorable environmental conditions as an important factor of the human environment; ensuring stability and the maintenance of a sustainable equilibrium of the state of ecological systems of the city; management of environmental processes in the big city; identification of the role of public authorities and local self-government in maintaining a healthy ecological environment of the metropolis; the development of the green economy as a mechanism modernization and innovative development of the territory; creation of the effective operational environmental monitoring in the conditions of a large city; increasing the level of ecological culture of the population and the extent of civil society participation in the formulation and implementation of environmental policy and other.

All of the above allows you to look at the city as a complex functional system in which an important place belongs to environmental problems. The spectrum of relevant issues covers the different aspects in the system "city - environment", which, in turn, allows fully and comprehensively review the formation of the environmental situation in the cities, to identify the arising problems and suggest possible ways of their solutions.

Every city is unique not only in its architecture, historical conditions of development, location and climate, but also on the emerging transport and economic relations and features of the production, including combinations of various industries, their composition, their branch specificity, the scale and nature of location, which is extremely important for the formation of the ecological situation in the territory.

The study of the ecological specificity of each city is the task extremely important because it determines the conditions of people's lives, their health, longevity and the comfort of the environment. This is particularly the relevant problem for large cities (Figure 1), because a large city - it, respectively, and the high concentration of population, industry, energy, transport, construction, etc. that leads to increased pressure on the environment and the formation of anthropogenic landscapes over large areas.

Figure 1. Possible consequences of impact of the large city on the environment
Source: composed by the author using (Burmatova, 2015a; Lvovich, 2006; Urbanization, 2011; Chubik, 2006; Matugina et al,, 2015) All this accompanied by a multifaceted impact on the environment, affecting all of its components - the atmosphere, hydrosphere, flora and fauna, soil, relief, climate, etc. As a result, in large cities, a new and largely artificial environment is created.

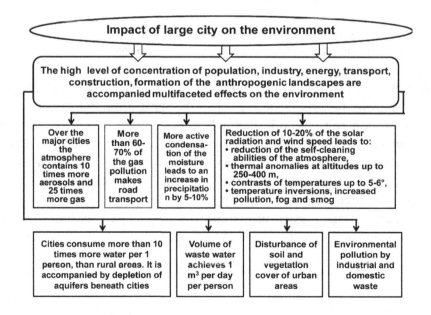

Place and Role of the Ecological Subsystem in the Structure of the Programs for Sustainable Development of Smart City

The concept of Smart City is based on the provisions of the sustainable development concept, the essence of which can be briefly reduced to the requirement of preserving and increasing the three potential - natural, technical and human, that is possible under the condition of simultaneous increase of the technical level of production, human health and the preservation of the proper quality of the natural environment, required to meet the needs of present and future generations. Thus, sustainable development is understood as the systematic unity of nature, economy and human, in the context of sustainable development. At the Summit in Rio de Janeiro ("Rio + 20") in 2012 the increasing significance of sustainable development has been demonstrated and acknowledged that ensuring the long-term successful development is possible only on the basis of the transition to "green" economy (Towards to a "green" economy, 2011). One way to achieve the goals of "green" economy is to ensure the environmental sustainability of cities (Resolution, 2015).

Ensuring sustainable environmental and economic development of cities requires joint consideration of the environment, production and population as elements of a single territorial-production system. One of the ways to reflect the environmental consequences of economic activity in the city is the development of ecological subsystem (or the environmental block) as part of its socio-economic development programs. Such subsystems ultimately encompass the main directions of the relationship of environmental, economic and social policies in the city. The result of the development of the ecological subsystem of socio-economic development of the city is to develop environmental management strategies, which can be regarded as a synthesis of its environmental problems and complex of nature protection activities aimed at their solution.

Strategic planning is an essential management function at various levels (including regional and local, including urban). It covers the process of selecting the city's development goals and ways to achieve them together into space (by performer), in time and finance. One of the elements strategic management tools are the environmental strategies and, accordingly, one of the subprograms as part of targeted programs of regions and cities are the subprograms for environmental protection. Environmental problems are usually long-term and require a strategic approach to their solution.

The logic of building a long-term environmental strategy as the landmark environmental policy is shown in Figure 2. As you can see, the development of environmental strategy involves a definition of the environmental mission of the city and its ecological image of the future, the choice of priority environmental goals and objectives, their detailing in particular for environmental programs, the

31

mechanism of their implementation of the production, analysis and evaluation of the results and consequences of programs. At the same time the environmental mission of the city and its environmental image of the future are based on the analysis of internal and external factors that influence the formation of the ecological situation, the establishment of the advantages and disadvantages in the field of environmental protection in the city, to identify opportunities and threats to the maintenance of environmental security of the urban economy and the environment of people. In what follows, we will adhere to the proposed logic of developing a strategy in Figure 2.

Consider the environmental subprograms in terms of their place in the structure of the regional programs of socio-economic development and to identify the relationship arising in the general system of the city, especially between the environmental, economic and social spheres.

The essence of the regional (municipal) programs as a form of implementation of the program-target method of planning, management and organization of social production, is to provide the solutions scientific, technical, economic and social problems of development of the territory, taking into account the specifics of its climatic, environmental and other conditions. In accordance with this, the regional program includes a set of measures to ensure a comprehensive and balanced (proportionate) solution of the problems of economic growth in a given period in

Figure 2. The logic of building a long-term environmental strategy as a landmark of environmental policy in a city
Source: composition of author

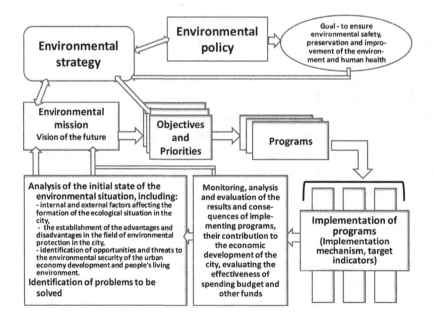

a particular area. Regional programs reflect, as a rule, the preferred embodiments of the socio-economic development of cities and strategic concepts of social and economic policy.

Sustainable Development Program for the city can be conventionally represented as a synthesis of the three main territorial subsystems (or blocks) - the economic, social and environmental (Figure 3). Each of them has its own goals and objectives, a set of measures, the composition of which depends on the resource and ecological potential of the territory, the current and future production and the spatial structure of the economy of the city, as well as character of the relationship between the respective subsystems. Let us dwell on the analysis of the place of the ecological subsystem among these elements of municipal programs.

Place of the environment subsystem in the structure of environmental strategy for sustainable development of the city is determined by the necessity of linking of the prospects of economic and social development of the city in compliance with certain environmental requirements. Accordingly, between the ecological subsystem and other key subsystems of the city - economic and social - are set close forward and backward linkages (Figure 3).

Relationships between the ecological and economic subsystems provide the exchange of the following information. From the economic subsystem it's possible to receive the information about a perspective production and the spatial structure of the city economy, on the basis of use of which in the environmental subsystem the compliance of the proposed economic development of the city to adopted the environmental requirements is verified. At the same time the environmental compatibility of various industries analyzed, the ability to Output above permitted standard pollution is determined, the potential violation of state of green plantations are assessed and etc.

The results of the analysis are the recommendations on the composition, arrangement, production scale, cost, etc., which allows to specify the possible production and the spatial structure of the economy taking into account environmental requirements.

From a social subsystem information on existing and forecast population, existing and future systems of settlement and level of development of social infrastructure goes. In the social subsystem requirements for the quality of the environment in terms of human health, the organization of recreational zones and tourism, the formation of the environment, etc. set (Figure 3).

Feedback from environmental to social subsystem provides for the establishment of the correspondence between the emerging environmental situation and adopted sanitary and medical requirements and on the basis of this - the formulation of proposals for the selection of possible modes of nature use in order to ensure favorable conditions for people's lives.

Figure 3. Place of ecological subsystem in the program of socio-economic development of the Smart City (relationships of economic, social and ecological subsystems city)
Notes: *The blocks marked by a dotted line city describe environmental subsystem. The blocks are located below the dotted line characterize the basic conditions of implementation of the environmental strategy of the city*
Source: *composition of author*

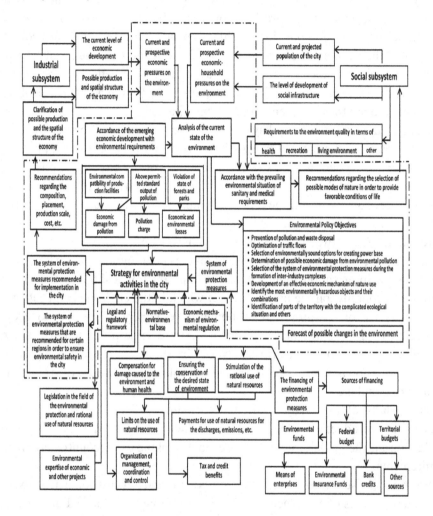

In Figure 3, in addition to links with the economic (especially industrial) and social subsystems, also logic of the developing the environmental subsystem, its main objectives and some of the most important conditions for the implementation are shown. The content of the environmental subsystem is heavily dependent on the intended directions and scales of development and distribution of productive forces within the study area, the formation of the territorial organization of the economy, the choice of settlement systems, etc. That is, there is a close relationship between the generation of environmental management strategies and mode of economic and social activity in the region.

There are three main elements in the environmental section, corresponding to the following research areas (Figures 3 and 4):

1. Analysis of the current state of the environment in the city;
2. Forecast possible changes in the state of the environment and the risk of ecological trouble;
3. Assessment of consequences of the impacts of human activities on the environment and developing proposals on the composition of environmental measures.

To assess the state of the environment in the city is needed, first of all carefully analyze the various factors that influence on the formation of the ecological situation. One of the important results of the analysis of the initial state of the environment in the city is the allocation of problematic situations (most environmentally hazardous sites and their combinations), and problem areas (parts of the territory with complicated environmental conditions). This is particularly relevant for the environmentally disadvantaged cities, because it allows you to identify "bottlenecks" with environmental positions and direct the forces and means in the first place on their unravel.

The following elements can be distinguished in the composition of the environmental subsistem:

1. Environmental facilities (wastewater treatment plants, equipment for the protection of the atmosphere, waste processing plants, and so on), for which in the process of solving problems is used to select treatment technology options, capacity, type, placement, taking into account their links with other elements of the economic complex of the city, as well as taking into account the environmental indicators considered production technologies;
2. Waste streams, reflecting the output of pollution in the environment; while the formation of the level of air pollution and water bodies should take into account all possible sources of emissions and discharges on the different types of hazardous substances and their combinations on condition the neutralization of pollutants emissions within acceptable limits;
3. Options for the main production technologies;
4. Arrangements for waste disposal of different types with the definition of volumes of their output, directions for use and distribution of recyclable products;

Figure 4. Scheme of formation of ecological subsystems in the system the environmental strategy of the city
Source: composition of author

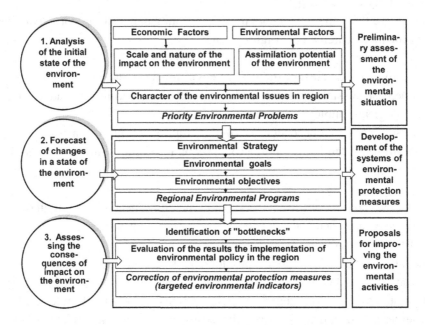

5. Measures to ensure conditions the wastewater discharge in open water objects, taking into account possibilities for water dilution, storage and natural movement of contaminants;

6. Costs for environmental purposes and economic damage caused by environmental pollution. The inclusion of environmental subsystem (environmental unit) in the measures of the regional programs provides, in particular, decision the following questions:

   a. Formation and the distribution of different types of waste (liquid, solid, gaseous);

   b. Adherence to accepted environmental standards;

   c. Air pollution, water and soil through the construction of sewage treatment plants and other facilities environmental infrastructure;

   d. Organization of recycling of generated waste;

   e. Consideration of a combination of pollutants in wastewater and gaseous and particulate emissions;

   f. Adherence to accepted environmental standards;

g. Air pollution, water and soil control through the construction of sewage treatment plants and other facilities environmental infrastructure;

h. Organization of recycling of generated waste;

i. Consideration of a combination of pollutants in wastewater and gaseous and particulate emissions;

j. Some specific conditions for the reproduction of water and land resources (including the conditions of formation and distribution of waste water, limitations on the possible volumes of their discharge, the conditions of the remediation work in the field of disturbed land, and others.).

In general, the environmental subsystem acts as a specialized tool for analyzing economic activity relationships and the environment in the city. Its specific composition is determined by specificity each city - its natural, economic and social conditions, the peculiarities of the territorial organization of the production and development prospects, as well as put forward by environmental objectives and nature of the problems to be solved.

The development and use of environmental subsystem of the urban programs allows not only analyze the interaction of elements of the economy, population and environment in the region, identify the necessary conservation measures, make a choice of options of developing and the distributing the production and population in its territory, under condition of fulfillment of the established environmental regulations, but also to determine the amount of costs required for environmental purposes, as well as the number of labor resources, the demand for local natural resources, infrastructure services needed to prevent negative environmental impacts of economic activities in the city.

Possible Directions of Predicting the Environmental Impacts of Economic Activity in the City

At forecasting the environmental consequences of economic activity in the city, there are the following methods of accounting for the impact of environmental factors on the elements of territorial and production systems (Figure 5).

The first way envisages carrying out the complex of nature protection measures aimed at the prevention, reduction or elimination of the negative impact of various objects in the environment.

Depending on the nature of the impact of an object on the environment, the possible negative effects of such an impact, the technical possibilities of their prevention, etc. implementation of environmental measures can go in three main directions.

Figure 5. The main directions of integration of environmental requirements in predicting the formation of the city's economy
Source: composition of author

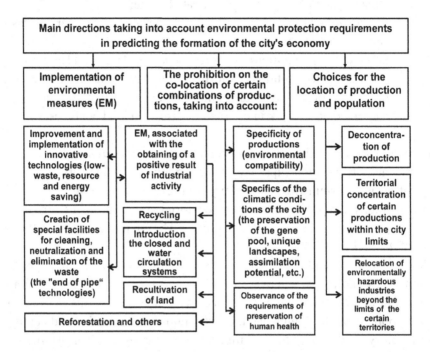

1. Improvement and introduction of innovative processes, which necessarily accompanied by improvement of production and environmental indicators (i.e. indicators characterizing the impact of technology on the environment and complete use of natural resources). This direction includes the expansion of the share of energy-saving and low-waste technologies. Reflecting such measures is possible primarily in the variants of functioning production facilities. At the same time improvement of technologies aimed at reducing their negative impact on the environment and applies both to the main production processes and on methods of waste treatment and environmental protection equipment.

2. Create special installations, equipment and facilities, other activities intended mainly for reducing pollution (the "end of pipe" technologies). Due mainly unidirectional action of this kind objects and events (to reduce pollution), they are called special-purpose. There are the sewage treatment plant for the treatment of industrial and domestic waste water, gas cleaning and dust removal equipment, Water-cooled plant and equipment, measures for dust suppression and dispersal of dust and gases, providing dilution of contaminated wastewater with fresh water, the establishment of buffer zones, gardening of

territory and others. Accounting for such measures when setting targets to minimize the negative impact on the environment is possible by selecting the appropriate option power structures, methods and techniques of neutralization of contaminants, cleaning efficiency indicators, the required expenses, etc.

3. Implementation of the various activities aimed not only at reducing certain types of human impact on the environment, but also connected with additional obtaining of a positive result of industrial activity (for example, issue of additional products, saving of resources, etc.). This is the so-called multi-purpose activities. They include waste management; implementation of sequential systems, recycling and closed water supply; land reclamation, the disturbed surface mining of minerals; reforestation and others.

A special place in the overall system of environmental protection measures belongs, in our opinion, to the organization of waste disposal, which is a necessary element in the overall chain of a system of non-waste productions. Greening of production is impossible without the addition of production facilities in special facilities designed for the processing of all kinds produced industrial and domestic waste. Disposal of waste, on the one hand, contributes to the reduction of allocation to the environment of various harmful substances, reduction of the scale of the negative impact of production on the state of the landscape, flora and fauna, etc. All of this eventually accompanied by a reduction of air pollution and water pollution, preservation of the landscape, the release of the territory as a result of the elimination of waste dumps, etc. On the other hand, waste disposal reflects the possibility of obtaining economic benefits - additional sources of release of a useful product, the possibility of expanding the raw material base of production through the use of waste and therefore saving natural resources, the ability to reduce the cost of the resulting utilization of waste products (through the use of cheaper raw materials, which is the waste), etc.

The second way of the integration of environmental factors when predicting the economic development of the area is the introduction of restrictions on co-location and functioning in a particular place different combinations of production, unfavorable from the viewpoint of sanitary conditions (Figure 5). The need for such a way is due primarily to the fact that, on the one hand, in the present available technical possibilities do not provide one hundred percent neutralization of waste and, on the other hand, focus on adherence to accepted environmental standards is not yet a reliable guarantee for the preservation of all the components of the environment (including human). Therefore, in the process of the implementation of environmental activities within the different areas additional measures of prohibitive character may be required.

A ban on the environmentally incompatible co-location facilities can be based on the following requirements.

Firstly, taking into account the peculiarities of enterprises considered in the limits of the city. Among these features should be called the peculiarity of the individual ingredients of emissions and discharges of different companies to interact with each other, leading to a deterioration of the quality of the environment or to the complication of the problem of waste disposal (effect of synergism). Harmful substances that have the effect of synergism, in conditions of joint presence may lead to the formation new compounds of heightened toxicity or enhancing the harmful effects of each other.

It is noted that the environmental compatibility of the plants largely depends on environmental danger of production technology, which is defined as by the physicochemical properties of the substances released into the environment, and their degree of the conversion and accumulation in the environment. A synergistic effect in the atmosphere possess, in particular, sulfur dioxide and chlorine, nitrogen oxides and hydrocarbons, sulfur dioxide and moisture, hydrogen fluoride and methyl mercaptan and others (Lystsov & Skotnikova, 1991; Petin & Synzynys, 1998). Synergistic effect of the influence in an aqueous environment have, in particular, ions of copper and lead (the presence of which combination of in sewage causes more damage than either of them separately), pesticides and oils, etc. (Combined (complex) action,n.d.).

Furthermore, there are persistent (non-degradable) contamination - substances and poisons, such as aluminum cans, mercury salts, phenolic compounds with long chain DDT, which in the natural environment or are not destroyed at all, or break down very slowly. For these substances, there is not natural processes, which could decompose them at the same rate as they are introduced into the ecosystem. Such non-destructive pollutants not only are accumulated, but often "biologically intensified" as far as passing in biogeochemical cycles and food chains. In addition, they can form other toxic substances, connecting with other substances surrounding environment.

Unfortunately, the account of the environmental compatibility of various industries belongs to the least studied and are usually ignored when taking economic decisions problems. At the same time, the interaction of pollutants entering the environment can lead to undesirable consequences for human health, the state of natural complexes and other recipients.

Secondly, when a ban on the co-location of certain industries need to consider not only the mentioned features of these industries, but particularly the territory of accommodation and its climatic conditions. This can be important when the target area has a unique landscape, valuable and rare species of plants and animals, etc. In this case, the decision to ban the co-operation of various industries in an area may be caused by requirements of the conservation of its natural complexes and a whole gene pool. Accounting for features of the area is of great importance and when the area is characterized by low assimilation capacity, i.e. low ability of ecosystems to neutralize harmful waste of human origin. This may be due to the unfavorable potential of climatic factors (associated with low air environment diffusing capacity), extreme climatic conditions (leading to a slowdown contamination decomposition processes), high already established load in the city on the soil, air and water (resulting in self-cleaning capacity different elements of the environment is already largely exhausted), etc. The presence of these conditions can lead to rapid and irreversible changes in the environment that requires particularly careful attention to such areas.

Thirdly, the establishment of restrictions on co-location and functioning of certain plants may be dictated by the requirements of preserving people's health (Characteristics of the main pollutants of the environment, n.d.). Different substances contained in wastewater and gaseous emissions, can have different impacts on public health. For example, nitrogen oxides cause a sharp irritation of the respiratory system; damaging effect on the respiratory system has and sulfur dioxide, which also causes an increase in susceptibility to infections, metabolic disorders, and others. In addition, under certain conditions the interaction of certain harmful substances may lead to extremely dangerous for human health compounds, for example, reacting benz(a) pyrene with nitrogen oxides leading to formation nitrabenzapirena which is highly active mutagen etc. (Burmatova, 2009). It is undesirable also the connection in one place for such activities which emissions in the air can lead to the formation of toxic or photochemical smog, which is also dangerous for human health.

The third way to integrate the protection of the environmental factor in predicting the formation of an economic complex of the city is associated with use of opportunities of the location of production and population to solve environmental problems (Figure 1.7). Such opportunities consist in primarily in the fact that considering a number of different variants of production operation (each of which is characterized by its environmental friendliness) it is possible to provide a choice of scale territorial concentration of production and the composition of such enterprises in each industrial site, that meet environmental requirements. The character of the placement and operation of production, the level of its territorial concentration and sectoral structure, existing settlement system largely determine the environmental situation in each place.

Thus, the environmental role of location and operation of production and population is mainly in the fact that with its help it is possible to regulate the concentration or dispersion within a single area of production and population in order to preserve the natural environment. At the same time the territorial dispersion of objects-pollutants makes better use of assimilation potential of the territory. Solving problems of protection of the environment by controlling the distribution of productive forces seems justified especially in cases where there are no sufficient technical and economic opportunities for the prevention, reduction or elimination of various adverse environmental impacts of economic activities. Therefore, this way of contributing to a partial solution of environmental problems, should be considered as one of the possible, and in some cases, necessary additions to other environmental practices, aimed primarily at the absolute reduction of harmful anthropogenic impact on the environment.

In general the forecast of possible changes in the state of the environment in the city is based on the account of economic development prospects within it, the current and expected production structure of the economy, the characteristics of sources of impact on the environment. In the case if production in the investigated territory is concentrated in certain parts of the city, it should be take into account that this entails, and increase of load on the environment in the relevant parts of the city and possible complications in their environmental situation. If the forecast the focus should be not so much to assess the impact of the individual objects in the environment, as the assessment of the total impact of territorial-production combinations on various elements of the environment, taking into account of background contamination, as well as other, already taking place, violations of the environment.

According to the results of evaluation of possible environmental impacts of the creation and functioning of the various objects the comprehensive forecast of the influence of the intended impact on the environment in the city are carried out. This forecast is directly related to the choice of the system of environmental protection measures, which should provide an approximation formed indicators of possible impacts on the environment (especially pollution) to the normative.

Prerequisites of the Development of Environmental Strategy

As already was mentioned, one of the most important tools of management the environmental sphere of the region and of forecasting its condition is strategic planning. The strategic planning process provides the basis for managing the city as a whole and its individual spheres, including, in particular, environmental. Development of a regional environmental strategy involves determining the environmental mission of the region and its ecological image of the future, the choice priority objectives and goals, detailing their in specific projects and programs, developing mechanisms for implementation, analysis and evaluation of results and consequences of implementing projects and programs. Consider on the example of Novosibirsk and Novosibirsk region[1] the main methodological aspects of the development of environmental strategy.

Environmental policies in the Novosibirsk region is aimed at maintaining the integrity of natural systems and providing a favorable environment for people. At the same time it's necessary to take into account that not only is a lot of accumulated in the past and so far unresolved environmental problems, but also the fact that the prospects of socio-economic development of the region associated with a possible further increase of the load on the environment. This, in turn, determines the relevance of development and implementation in the medium and long term, an adequate system of environmental measures that could form the basis for the strategy of environmental activities in the region and, consequently, to determine the main directions of improving the environmental regulatory system in this region that minimize the negative impacts of human activities on the environment and in general of ecological safety of the economic development of the Novosibirsk region.

The basis of the system of prediction of the environmental situation in the Novosibirsk region is the development and implementation of a number of interconnected documents, including the following elements:

1. The section "Ensuring environmental safety and environmental protection" as part of the Strategy for Socio-Economic Development of the Novosibirsk region for the period up to 2025 (Strategy, 2009);
2. Long-term target program "Environment of the Novosibirsk Region" within the Program of the Socio-Economic Development of the Novosibirsk region up to 2015 (The Program, 2009 The Concept, 2009);
3. Strategic Plan for Sustainable Development of the city of Novosibirsk and complex target programs. - Novosibirsk, 2004. - 245 p. (Strategic Plan, 2004).
4. Subsection "Solving the environmental problems" of the Strategy of socio-economic development of Siberia up 2020 (Section IV. Priority Interbranch development directions of Siberia) (Strategy of socio-economic development of Siberia, 2007; Economy, 2009).

It is important to note that forward-looking strategic developments in the field of environmental protection are part of the general system of strategic planning documents of socio-economic development in the region (Economy, 2009; Kuleshov& Seliverstov, 2016; Seliverstov, 2012; Seliverstov & Melnikova, 2013) and have a close connections with economic and social issues in the region, as oriented on a comprehensive solution economic, social and environmental problems. In accordance with this, the formation of long-term target program "Environmental Protection Novosibirsk Region" (hereinafter - the Program) suggests as a first phase the elaboration of the project concept of this program, which are built with taking into account the connection with the Program of socio-economic development of the Novosibirsk region in 2025 (Strategy, 2007; The program, 2009; Seliverstov, 2012). The concept of long-term program "Environmental Protection the Novosibirsk region" is intended to further the formation of the long-term program, as well as definitions of the main contours of the internal structure of the section dedicated to environmental protection, its place in the program of socio-economic development of the Novosibirsk region and the total system of strategic planning area.

Development and implementation of the Program should begin with an analysis of the initial state of the environment in the region and should end with the elaboration of defined measures for the desired adjustments of the nature management and forming a healthy environment within the territory under conditions of the quality control of the entire process of development and implementation of the Program, the necessary coordination in the relevant activities and evaluation of the results (Figure 6)[2].

Factors Determining the Formation of Environmental Situation in the City

Environmental policy in the region or city is largely dependent on how adequately in its implementation taken into account factors affecting the development of the environmental situation. Analysis of these factors in relation to the possibility of the environmental governance allows you to identify the most significant for a specific

Figure 6. Structure and scheme of the process of the development and implementation of the Program "Environmental Protection the Novosibirsk region"
Source: composition of author

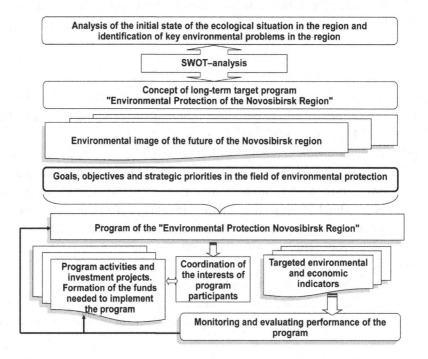

period of time environmental problems and offer the ways for their solutions. It is important to be clear about whether the critical factors amenable to control by the environmental agencies, whether they are internal or external, within the scope of regulation of environmental field or external conditions, on which directly affect the regional authorities are not able.

In every city the environmental situation, priority environmental issues and finding approaches to their solution are determined, as a rule, by three groups of aggregated factors:

1. The specific conditions of the region or city (spatial aspect);
2. Industry features productions that are presented in the region or in the city (production aspect);
3. Approaches to solving environmental problems in the framework:
   a. Separate economic entities (organizational and technological aspects),
   b. The city as a whole, including the environmental policy at the level of the city and its districts (management aspects).

In more detail, these factors are presented in Figure 7.

Results of imposition of these factors on the territory of the Novosibirsk city (Figure 8) testify that from the standpoint of the formation conditions of the environmental situation, the region relatively lucky only with branch structure of production (meaning first of all the absence of typical polluting industries). For the remaining factors the situation is sufficiently urgent that in the perspective of socio-economic development of the region under study determines the urgency of developing and implementing the necessary environmental measures that form the basis of the program in the field of environmental protection and determine the main directions of improving the system of environmental management in the region.

Features of the environmental situation in the Novosibirsk region and emerging environmental problems are mainly caused by local climatic conditions and character of influence on them the region's economy (industry, energy, transportation, utilities and agriculture), which in turn depends largely on the specific location of industrial enterprises, their capacities, technologies used, the extent of the territorial concentration of production and population, the existing level of a violation of the natural environment in the region and other conditions. Concretization of these conditions with respect to the considered region shown in Figure 8.

Projected in the Novosibirsk region economic development involves the growth of the fuel industry, ferrous and non-ferrous metallurgy, chemical industries, construction materials industry (cement) and freight turnover transport (Strategy, 2007). This can lead to increased pressure on all parts of the environment that will require conducting adequate environmental measures.

Figure 7. Factors influencing the choice of environmental situation on a territory
Source: composition of author

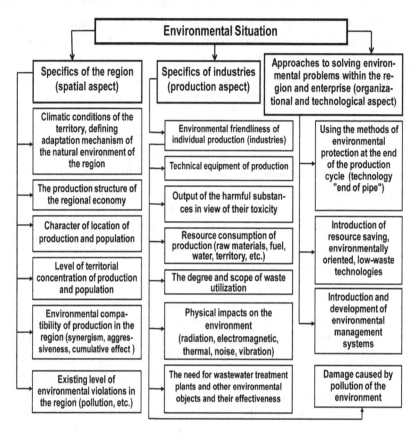

Analysis of the Initial State of the Environment and the Major Problems

A high concentration of industrial production and population In Novosibirsk causes an increased flow of pollutants into the environment of the city, posing a threat to human health. Novosibirsk is the third city in Russia by the number of residents in the population (1.6 million. Pers.). Environmental situation in Novosibirsk, as well as other big cities, is dependent not only on the released into the air of harmful substances, but also on a variety of adverse weather factors, such as calm, temperature inversions and fog (which are conducive to the accumulation of harmful substances in the surface layer of the atmosphere). According to studies of Roshydromet[3], Novosibirsk is located in the area of high potential of air pollution[4] (PAP), i.e. in the zone of unfavorable meteorological conditions for the dispersion of contaminants, so that in some periods of intense accumulation of harmful substances can occur

Figure 8. Features of the Novosibirsk from positions forming the ecological situation (in italics are marked favorable factors)
Source: composition of author

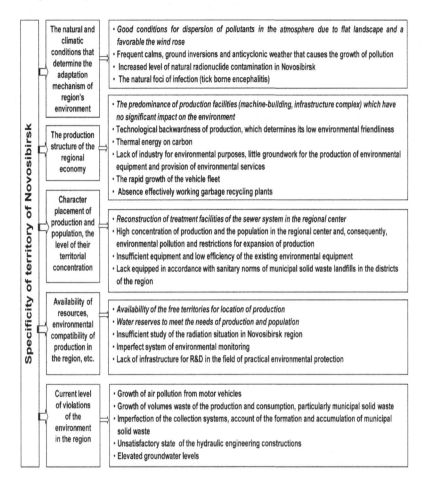

in the atmosphere and the formation of high levels of pollution. In general, the scattering ability of the atmosphere in Novosibirsk much better than, for example, in Eastern Siberia, or Kuzbass (the Kemerovo region), but they still do not reach the appropriate level, which is observed in the European part of Russia, for this reason, the city has increased the meteorological pollution potential[5].

Main indicators characterizing the state of the environment in the territory of the Novosibirsk region from positions of atmospheric air, water basins and waste in the dynamics are shown in Table 1 (Burmatova, 2015a; State reports, 2015; State report, 2016a).

Despite the growth in industrial production and an increase in the number of vehicles, air quality in large settlements of the Novosibirsk region (including Novosibirsk) in recent years has remained relatively stable (Burmatova, 2015a; State reports, 2015; State report, 2016a; The State Report, 2016b) (Table 1 and Figures 9 and 10).

Notes:

1. The table is compiled using data of the Novosibirskstat and the Department of Natural Resources and Environmental Protection of the Novosibirsk region[6], as well as sources (Burmatova, 2015a; State reports, 2015; State report, 2016a; The State Report, 2016b).
2. Calculations are made by the Department of Natural Resources and Environmental Protection of the Novosibirsk region under the simplified procedure developed by "SRI Atmosphere", using specific emission factors.
3. In the denominator is indicated polluted waste water without treatment (in total volume of wastewater discharges).

Table 1. Main indicators characterizing the impact of economic activity on the environment of the Novosibirsk region[1)]

Indicators	2009	2010	2011	2012	2013	2014	2015
Extraction of water from natural water bodies - total, million m³, including:	755,2	763,6	676,0	703,9	646,6	634,8	639,73
- from surface water bodies	659,5	696,5	613,4	642,8	586,19	574.83	579,59
- from underground sources	95,7	67,1	62,6	61,1	60,4	59,97	60,14
Wastewater discharges into surface water bodies total, million m³, including:	588,2	604,2	527,1	544,2	512,16	500,6	491,7
- discharge of polluted wastewater, million м³	98,4	106/ 48,3[3)]	92/ 34,9[3)]	112,5/ 40,1[3)]	114,7/ 33,96	109,35/ 33,93	106,94/ 30,99
- discharge normatively treated wastewater, million m³	254,1	280,18	231,77	248,23	222,0	220,28	216,19
Air emissions, thousand tons: - from stationary sources	233,5	228,4	234,0	224,5	195,7	207,8	184,7
- from motor vehicles	358,4	319,9	287,4	286,2	310,2	276,5	275,2
Pollutant emissions from stationary sources, per 1 inhabitant, kg	88,1	85,7	87,1	83	71,65	75,7	66,9
Waste generation of production and consumption, million tons	1,91	2,07	2,5	1,8	1,863	1,95	3,89

For the past 7 years, emissions of air pollutants from stationary sources per capita decreased from 88.1 to 66.9 kg (Figure 10).

Table 2 lists the most important ecological problems of Novosibirsk that require focus and solution in the mid- and long-term perspective, as well as possible ways of solving them.

Environmental Vision of the Future of the Region, the Challenges in Creating a Favorable Ecological Situation, Strategic Environmental Goals, Objectives and Priorities

By analyzing the ecological situation in Novosibirsk and Novosibirsk region, it is possible to form an ecological vision of this region's future, delineate challenges in creating a favorable ecological situation, and determine the strategic ecological priorities and directions of nature conservation activity.

The main elements of the vision of this region's future are characteristics that presuppose improvement in the quality of the environment and ecological conditions for human vitality, including the creation of a healthy environment, ecologized industry, organization of an efficient ecological sector of the economy, and nature conservation and protection (Figure 11). The ecological vision of the future (which is, however, in the territorial aspect sufficiently universal, since it is based on common social, economic, and ecological requirements) envisages an aligned foundation on

Figure 9. Dynamics of emissions of harmful substances into the atmosphere of the Novosibirsk region in 2009-2015 (thous. tonnes / year)

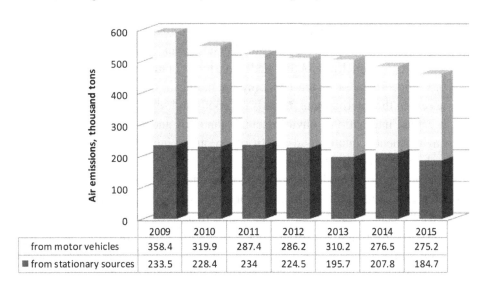

	2009	2010	2011	2012	2013	2014	2015
from motor vehicles	358.4	319.9	287.4	286.2	310.2	276.5	275.2
from stationary sources	233.5	228.4	234	224.5	195.7	207.8	184.7

Figure 10. Change of air emissions of pollutants from stationary sources (kg / per 1 person) on years

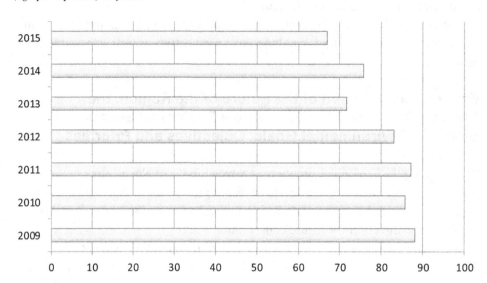

which the necessary conditions should be created to make the harmonic interaction of the regional economy and ecology to become a reality.

In modern conditions of economic development the maintenance of the ecological safety in Russia faces serious challenges and threats which require for their solution the participation of the authorities at all levels of management.

The main challenges in creating a favorable environment in the regions of the country, including the Novosibirsk region are the following circumstances.

1. The need for transition to a new life and environmental safety standards, the urgency of the introduction of resource-saving and environmentally friendly technologies which are conditioned by the needs of modernization and increase of competitiveness of the economy, including tightening of conditions for access to the international markets for goods and services in accordance with accepted international environmental standards and increasing demands for environmental quality and safety of products, including in environmental technology parameters.
2. The possibility of modernizing the economy during the economic crisis, which involves the need for the introduction of energy-saving and environmentally friendly technologies and to increase public funding of environmental measures. This, in turn, requires a strengthening of environmental priorities in the state economic and social policies that will eventually lead to an increase in the competitiveness of Russian companies on the world markets.

Table 2. Most important ecological problems in Novosibirsk and possible ways of solving them

Problem	Possible Ways to Solve
1. Air pollution by stationary emission sources	• Implementation of ecologically oriented technologies at businesses of the city, both active and planned; modernization of equipment at a number of active businesses. • Continued operations to supply gas to communal and government-office furnaces; closing of ecologically unfavorable furnaces (the number of which in Novosibirsk exceeds more than 100 objects). • Improving control of atmospheric emissions, primarily in the districts of the city where concentrations of harmful admixtures in the air regularly exceed the maximum permissible concentration. • Greening of urban territories (tree planting, increasing urban forest area). • Greening of urban territories (tree planting, increasing urban forest area). • Improving the fining system for negative anthropogenic impact on air quality. • Development for city objects of regulations on maximum permissible emissions of harmful substances into the air. • Revealing of problem areas of air pollution in the city
2. Air pollution from mobile emissions sources (auto transport)	• Road repairs and construction of new transport interchanges and bypasses in Novosibirsk. • Ensuring conditions for rapid transfer of part of the auto fleet (including municipal transport) to motor fuel. • Organization of industry and implementation of auto exhaust neutralizers. • Strengthening of state control over compliance with norms for permissible smog and harmful substance emissions by auto transport, which accounts for around 70% of the total harmful emissions in the city. • Optimization of road traffic, ordering of transport flows in city, determination of places where auto transport is concentrated in populated areas. • Improving the fining system for environmental pollution from mobile sources.
3. Organization of rational use and conservation of water resources	• Creation of combined water supply systems (using surface and underground water). • Reconstruction of sewer and water main networks. • Modernization of treatment facilities at active businesses of the city. • Measures on reducing volumes of dumping polluted sewage into surface waterbodies. • Implementation of water recycling systems. • Organization of repeat (sequential) use of treated waste water between businesses of city. • Improving drainage sewers in city. • Accountability of ecocompatibility of different industries from the standpoint of water pollution and organization of where to locate water discharge points. • Placement of new industries with allowance for requirements of waterbody conservation. • Improving the payment system for pollution of waterbodies. • Improving the system for monitoring the water basin, expansion of a continuously acting observation network for the state of surface waterbodies. • Revealing of "narrow" areas in the city from the standpoint of pollution of waterbodies. • Continuation of operations on restoring, cleaning, and conserving small rivers bordering Novosibirsk and other cities of the region.
4. Improving the situation in dealing with industrial and consumer waste	• Construction of areas of solid household waste in Novosibirsk, their development using modern technology. • Construction of new waste treatment plants. • Organization of solid waste sorting. • Elimination of unsanctioned dumps. • Utilization of thermal energy from ash and slag created at factories by using it as raw materials at construction industry entities. • Collection and utilization of household waste from the private sector with full coverage of private homes. • Creation of industrial businesses for secondary waste recycling (including paper).
5. Ensuring soil protection	• Measures on combating wind and water erosion of soils. • Compensation for losses from confiscation of agricultural holdings for construction. • Recultivation of damaged lands.
6. Improving the situation with radiation	• Improving and developing a system for observing and controlling radiation in the region. • Monitoring of the level of impact of natural radon on the inhabitants of Novosibirsk. • Testing of the territory of Novosibirsk for radon hazard.
7. Ensuring reproduction of fish stocks	• Measures on conserving and raising commercial fish species (the Ob River and the Ob reservoir). • Measures on preventing and compensating damage to the fishing industry in connection with the destruction of spawning grounds and the death of young fish and egg. • Intensification of combat with poaching in rivers and lakes of the city and region.

continued on following page

Table 2. Continued

Problem	Possible Ways to Solve
8. Reduction in the water level of the Ob River and Ob reservoir, progress-ive underflooding the territory of city	• Measures on ensuring safe functioning of hydroengineering installations • Fortification of the shoreline of the Novosibirsk reservoir, improvement of water conservation zones in the Ob River basin. • Improvement of urban drainage networks
9. Development of an environmental monitoring system	• Creating a network of operational monitoring and control of the environmental situation
10. Creating a unified system of financing environmental protection measures	• Attraction of investments for nature conservation. • Use of budgetary means of different levels. • Strengthening of the role of the oblast budget in financing ecological programs and measures. • Introduction of mandatory ecological insurance of a number of potentially hazardous industries and technologies.
11. Education in ecological culture	• Involvement of society in preparing and implementing ecologically important solutions. • Creation of centers for ecological education of the population

Source: The table is compiled using data of the Department of Natural Resources and Environmental Protection of the Novosibirsk region[7], as well as sources Burmatova, 2015a; State reports, 2015; State report, 2016a; The State Report, 2016b).

Figure 11. Environmental vision of future for Novosibirsk and Novosibirsk region
Source: comosition of author

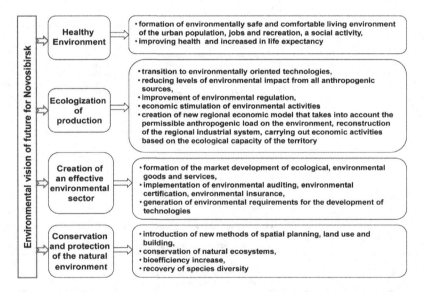

3. The weakness of the current state of environmental policy in the Russian Federation, which is characterized in particular by the following main features:

    a. the imperfection of assessment of the taxable base of the using natural resources (low interest rate payments for the use and reproduction of natural resources; the practical absence of the payments for recycling and re-side use of natural resources), etc.;

    b.   low specific weight of resource taxes and environmental payments in the payment system, forming the budgets of all levels;

    c.   insufficient financing environmental activities from the budgets of all levels, specialized funds, funds of enterprises and etc. One result of inefficient state environmental policy is unsatisfactory state of the environment in many regions of the country;

    d.   existing practice of levying payments for negative impact on the environment has a number of weaknesses:

        i.   are extremely low base rates, which is why their role is purely symbolic;

        ii.   the legal status of these payments is not adjusted, the relevant law is not adopted so far, although it stems from the need of the law "On Environmental Protection (as amended December 29, 2015)";

        iii.   refusal of targeted use of the relevant funds in the budgets of the Russian Federation and the regions;

        iv.   a set of substances for which payments have been established, is far from complete;

        v.   insufficient consideration of inflation factor - adjustment of payments occurs with a delay, and the value of the correction factor to the regulatory charge is not comparable with the actual rate of inflation.

4.   The need to develop and implement new and effective tools in the field of environmental regulation, allowing to stimulate, on the one hand, the ecological modernization of production, development and use of environmental technologies, the formation of the market environment-friendly products and environmental services; and, on the other hand - an environmentally responsible behavior of business. Such tools should be supported by both legislation and provided with appropriate implementation mechanisms.

5.   Technological backlog of industrial enterprises (for example, most of the technology engineering companies refer to the 4th technological way), physical obsolescence of technological equipment park that is directly related to the environmental risk and economic inefficiency of the technology used, and requires a transition to new environmentally safe technologies.

6.   The need for improvement and strict compliance with environmental legislation, which, despite all the positive intentions, remains almost unchanged. Emasculation of the essence of environmental legislation as a result of numerous reforms has led, in particular, to the fact that the environmental impact assessment of projects (including the environmentally dangerous) turned out to be unnecessary, resulting in the growth of the environmental risks and is fraught with unpredictable consequences for the safety of the environment and human health. The main drawback of the existing environmental legislation is

the absence of it instruments to promote the use of environmentally friendly technologies. As a result, objects-pollutants is more profitable to pay fines for environmental pollution (because their level is still quite low), than to carry out the actual environmental activities.

7.   Imperfect methods to determine the economic damage caused by economy and people's health by pollution of the environment; the need to eliminate accumulated environmental damage from past economic activity.

8.   The imperfection of statistical reporting on the use of natural resources and environmental protection, weak control financial discipline of resource and environmental charges; the absence of a modern system of monitoring, evaluation and prediction of the environment, emergency situations of natural and man-made disasters and adverse climate changes.

9.   The tightening of conditions for access to international markets from the standpoint of environmental standards and regulation; increased international competition due to increasing demands for environmental quality and safety of products, the transition to the integration of environmental parameters, not only products, but also the technology used for its production.

10.   Low environmental responsibility of the business, which is manifested especially in the weak economic incentives for usingf natural resources to comply with environmental requirements, and as a result - low investment activity in environmental protection measures. Ignoring the ethical aspects in the field of ecology and generally low ecological culture of the population. Ecological awareness in Russia is at the initial stage of formation, so it is extremely important not only the formation of environmental ethics, a respectful attitude to natural environment, the strengthening of the principles of eco-efficiency and environmental justice, but also timely ethical evaluation by the country's leadership the facts of gross violation of the environment.

11.   The need for a purposeful state economic policy focused on the greening of production, provided a systematic approach to solving problems of structural and technological transformation of the economy in favor of resource-saving and environmentally friendly industries that would not only lay the foundations and to strengthen the innovation economy, but also to provide both economic and environmental benefits.

12.   The need for the use of strategic planning and management in the formation of the state environmental policy, since environmental problems are usually long-term and require a strategic approach to their solution.

13.   Nature conservation activity in Novosibirsk consists of measures aimed at solving revealed and formulated ecological problems that have both occurred on its territory and may occur when implementing planned investment projects,

as well as of a system of nature conservation measures determined by the need to alleviate or warn about possible ecological problems (see Table 2).

In accordance with this logic, let us consider the main steps in developing an environmental protection strategy in the region (Fig. 12).

The strategic role of the regional ecological program can be formulated as providing ecological safety in Novosibirsk region by stabilizing and improving the ecological situation, and conserving and restoring the integrity of natural ecosystems. This aim can be achieved by solving the following problems:

- Conservation and improvement of the quality of the natural environment, reduction of the negative impact on it during scientific and technological developments of the economy;
- Ensuring protection of the population, objects of the economy and territory of the city and region from harmful water impact;
- The creation of an effective system for dealing with industrial and consumer waste;
- Protection of the city from radiation, as well as reduction to a socially acceptable level of the risk of radiation impact on people and the environment they inhabit;

Figure 12. Scheme of the process of development and implementation of the Program "Environmental Protection Novosibirsk Region"
Source: composition of author

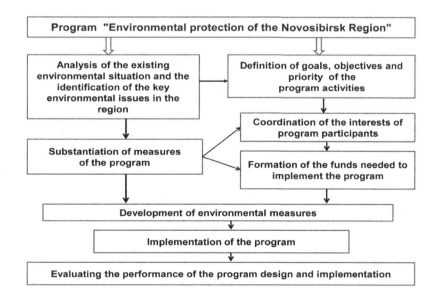

- Implementation of measures for preserving and reproducing (restoring) renewable natural resources as components of the natural environment;
- Observance of ecological regulations on economic activity envisaging correspondence of its scales and risk of impact to the ecocapacity of natural systems;
- Development of a network of specially protected natural territories and conversation of biodiversity;
- Improvement in the system for regulating natural resource use, environmental conservation, and the corresponding regulation mechanisms;
- Increasing the level of ecology education and advocating care for nature.

In accordance with the stated aims and problems of the program "Environmental Conservation of Novosibirsk region" and the main positions of the Socio-economic Development Strategy of Novosibirsk region for the Period up to 2025, the overall strategic priorities for environmental conservation are as follows:

1. Wide use of ecologically clean technologies (at both active and planned plants) to provide a technological basis for ecologically safe economic activity with transfer of the regional and state economy on the whole to an innovative path of development;
2. The equipping of businesses with nature conservation equipment, technological reequipping and gradual retirement of businesses with obsolete equipment, reduction in total water consumption in production and communal housing, and development of a system for using secondary resources, including waste recycling.
3. Ecological regulation of economic activity that bring into correspondence with established ecological standards the scales of the latter and the impact risk of individual industrial objects (primarily the fuel energy, metallurgical, and chemical complexes as the ecologically most important sectors of the oblast's economy) and territorial concentration of production in individual parts of the region;
4. Elaboration and implementation of measures to strengthen territorial administrative structures active in environmental conservation, the development of an ecological monitoring system, expansion of ecological control of industry and types of activity, including potentially hazardous types, independent of administrative jurisdiction and forms of property;
5. Prediction of the level of ecological expenditures for the oblast as a whole and for individual elements of its economic complex with allowance for planned production growth rates;

6.   An ecological accountability mechanism for the economic activity of subjects for negative impact on the environment and increased interest in implementing nature conservation activity, including improvement in the order of fines for negative impact on the environment.

In accordance with these ecological priorities, nature conservation activity in Novosibirsk should be directed at the following: sequential reduction in negative technogenic impact on the environment; prevention of degradation of natural complexes when implementing new investment projects; use of a nature factor for restoring and improving people's health (primarily by development of tourism and creation of recreation zones); increased quality of the drinking water supply; combating underflooding of the territory and feral herd infection (tick-borne encephalitis).

Orientation toward these priorities will make it possible to gradually improve environmental quality in Novosibirsk and, on this basis, implement principles of stable development of the oblast in the mid- and long-term perspective with allowance for solving problems on environmental conservation.

Targeted Environmental Indicators

To assess the effectiveness of the ecological policy in the region from the standpoint of achieving the stated aims and tasks, a system of indices can be used that characterize the ecological processes occurring here and include the entire set of parameters that adequately describe the state of the natural environment. The system of indices were substantiated previously and published by us in Burmatova (2015b).

The tasks of assessing the ecological and economic effectiveness of the measures of environmental conservation of Novosibirsk region are as follows: to obtain quantitative criteria on the admissibility or inadmissibility of implementing a particular measure; to ensure a choice of planned economic activity with the least losses; to obtain quantitative criteria on assessing the effectiveness of planned nature conservation measures.

The main generalizing indices of the efficiency of carrying out these measures can be as follows:

1.   The total amount of emissions of pollutants into the environment (atmosphere and water basin, in tons and cubic meters, respectively, per person per year);
2.   A reduction in the volume of air pollution from stationary sources per unit of gross regional product (GRP);
3.   A reduction in the volume of air pollution from mobile sources per unit of GRP;

4.   The mean annual growth (reduction) of the volume of pollutants dumped into water bodies per unit of GRP;

5.   A decrease in the volume of unrecycled industrial and consumer waste.

A reduction in pollutants into the environment per unit of GRP means a strengthening in the ecocompatibility of the applied technologies, growth in the efficient operation of gas-purification equipment, a reduction in the energy intensity of production, improvement in the quality of the environment, and a decrease in the negative impact of the economy on health. Increasing the level of recycling and rendering wastes safe will entail a reduction in the ecological hazard of waste accumulation and will serve as a characteristic of the effectiveness of the waste regulation system. The achievement of planned targeted environmental targets would indicate the possibility of not only preserving the achieved quality of the environment, but also its progressive improvement.

In addition, to assess the results of achieving the aims and solving the tasks of the program, specific particular indices can be used that characterize different aspects of the impact on the region's environment taking into account its economic, social, and ecological specifics.

To solve the stated problems, it is necessary to have the targeted ecological guidelines—indices for the effectiveness of environmental quality control, which are listed in Table 3 (prognostic estimates have been made using a mobilization development scenario for Novosibirsk region (Strategy, 2007)).

The air pollution forecast for Novosibirsk in terms of the most widespread pollutants coming from stationary sources for the period of 2005–2025, taking into account the aforementioned measures and parameters of the mobilization scenario (Strategy, 2007), testify the real possibility of a trend toward gradual improvement in air quality in the region (Figure 13).

Table 3.Forecasted target environmental indicators for the Novosibirsk region

Indicators	2010-2013	2014-20016	2017-2020
Average annual increase in the volume of emissions into the atmospheric air from stationary sources per unit of the GRP	8-10	7-9	5-7
Average annual increase in the volume of emissions into the air from mobile sources per unit of the GRP	13-15	12-14	10-12
Average annual increase in the volume of waste water discharged into water bodies per unit the GRP	16-18	12-14	9-11
Volume of reduction of the unprocessed waste production and consumption, thousand tons	120–150	160–200	210–250

Source: calculations of author

Figure 13. Forecast of air pollution by the major pollutants by 2025
Source: *calculations of author*

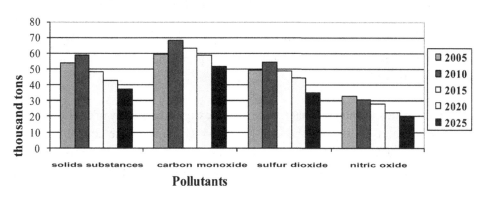

The target ecological indices reflecting the strategic ecological priorities and key ecological problems are designed to be used by regional authorities as a tool for monitoring activity in the sphere of ensuring stable territorial growth. This makes it possible to simultaneously improve and adjust the measures of the program along with ensuring that the ecological goals in the region's development are met in case this activity is ineffective.

Meanwhile, these indices should be accessible and clear to other interested stakeholders, including social organizations and the regional population. This is a very important condition, because ensuring the ecological welfare depends equally on the actions of individual people and on the work carried out by regional authorities. The corresponding structural administrative subdivisions of a federal subject collect data and prepare reports on the actual values of the indices for stable development so that all interested parties can see how effectively the program has been implemented.

One of the most important conditions for effective implementation of strategic ecological developments in the region is strengthening of state regulation in environmental conservation. The particular urgency in relation to this assumes the creation of an economic mechanism of ecological regulation aimed at stimulating rational natural resource use and environmental conservation; spurring and supporting ecologically responsible businesses; and improving the administrative regulatory structure and legal bases of environmental conservation. In many aspects, the effectiveness of implementing mid- and long-term prognostic documents concerning ecology also depends on the quality of the corresponding mechanism.

Using a Program-Oriented Approach, the Possible Variants of Solutions of the Problems and Assessment of Risks

Integrated solution of the mentioned problems of natural resource management and environmental protection in Novosibirsk requires the use of program-oriented approach, which allows you:

1. To take into account the magnitude, complexity and diversity of environmental problems of the region, the solution of which requires the consolidation of efforts and funding sources for the development and implementation of complex interrelated specific tasks, resources, and implementing activities of different nature to achieve the goals;
2. To coordinate the goals and objectives of the Program with goals and objectives of the other long-term programs of the Novosibirsk region ("Development and distribution of productive forces the Novosibirsk region", "Development of innovative activity in the economy and the social sphere in the Novosibirsk Region", "Development of transport infrastructure Novosibirsk Region", "Energy development, energy efficiency and energy security of the Novosibirsk region", etc.);
3. To combine the administrative and control tools of management and market economic principles, thus ensuring the coordination of various aspects of environmental activities of businesses;
4. Consistently integrate environmental objectives into the process of the socio-economic development of the area in order to ensure sustainable development;
5. To ensure the harmonization of setting and achieving balanced current and long-term environmental objectives;
6. To establish a clear priority in meeting the investment needs in the field of environmental protection, given the limited resources.

Thus, the program "Environment of the Novosibirsk Region" is seen as a key tool for planning, forecasting and implementation of regional environmental policies, as well as coordination of environmental activities in the Novosibirsk region. At the same time, the program acts as a method of implementation of the Strategy of socio-economic development of the Novosibirsk region for the period until 2025 and the Socio-Economic Development of the Novosibirsk region for the period until 2020.

The task can be solved by the several options corresponding mobilization (baseline) scenario Strategy of socio- economic development of the Novosibirsk region until 2025 (Strategy, 2007). Under this scenario, we consider two possibilities, reflecting the minimum and maximum options of mobilization scenario. Minimum scenario simulates the development of the Novosibirsk region on inertial type (reproducing the conditions and restrictions of 2000-2005). Maximum scenario describes the most complete use of the basic potential conditions for the development of the Novosibirsk region. Accordingly, there are two possible options for solving environmental problems.

The first option largely reflects current trends of development and distribution of productive forces of the Novosibirsk region and provides for the implementation approach to solving environmental problems, including mainly the establishment and improvement of methods and means to protect the environment at the end of the production cycle. In this case, the completion (or addition) of existing fixed manufacturing technologies or individual objects by different systems disposal of waste is carried out in order to prevent certain scale negative impacts on the environment (including treatment facilities for treatment of contaminated wastewater, installations for dust and gas extraction, water recycling system, the organization of waste management, construction of waste treatment plants, etc.) as well as activities to eliminate negative already committed violations in the state of the environment.

However, the possibility of the first variant, first and foremost in terms of economic and environmental efficiency of the technologies "end of pipe" are rather limited mainly due to the difficulties in achieving sufficiently high degree of purification of emissions and discharges, as well as in connection with an exponential relationship between the degree of extraction of harmful substances contaminants and level of expenditure for the necessary environmental protection measures.

At the same time, the failure to comply with of measures of this option significantly increases the risk of harm to the environment and human health. In addition, the cost of rehabilitation of territory in the event of violations (especially pollution) greatly exceeds the amount of investment required to prevent such violations.

In the capacity of the ecological risk assessments can be used indicators characterizing:

1. An increase in air pollution and acid rain formation (due to huge amounts of emissions of sulfur dioxide and nitrogen oxides produced during combustion, acid rain reduces the crop, destroys vegetation, destroys life in fresh water, destroys buildings, increases corrosion of metals and etc.);
2. Changes in the qualitative and quantitative status of surface and underground water sources (under the influence of excess pollution, violations of the hydrological regime of rivers caused by different kinds of human impact, etc.);
3. Formation of hazardous waste (toxic and radioactive) above permissible limits;
4. The expected economic damage (calculated and prevented) from possible contamination of the environment (air, water, soil, mineral wealth, etc.);

5. The volume of greenhouse gases (carbon dioxide, nitrogen oxides, methane, chlorine, etc.) and their accumulation in the atmosphere above certain concentrations (established the environmental standards for relevant ingredients);

6. Emissions of ozone-depleting substances (CFCs, chlorine and its compounds with oxygen, greenhouse gases); known that reducing the ozone layer by 1% leads to an increase of ultraviolet radiation by 1.5% and a corresponding increase in skin cancer from 2 - 3 to 5-7%. In addition, fall harvest crops, reduced phytoplankton productivity, the loss of many species of fish and marine invertebrates, etc. take place.

The procedure for evaluating the likely environmental risk is shown in Figure 14.

In the case of the using the second variant of solution to the problem in addition to the measures of the first variant is provided modernization and technical re-equipment of production through the introduction of resource-saving and low-waste technologies. This variant is characterized by a high economic and environmental performance compared to the first, it will allow qualitatively change the ecological situation in the region through technical upgrading of existing facilities and the introduction of new facilities on the basis of high technologies for environmental safety operation of enterprises.

Currently, however, the second solution to the problem seems to be premature in view of the fact that in modern Russian conditions, exacerbated by the effects of

Figure 14. Procedure for evaluating the environmental risk
Source: composition of author

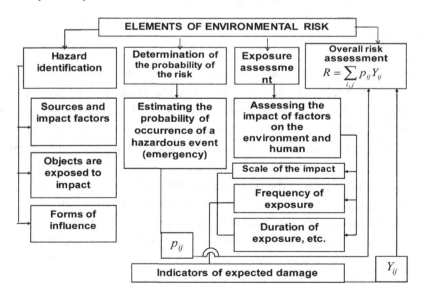

a deep global economic and financial crisis the implementation of such large-scale, technically complex and expensive programs without preparatory activities and to develop mechanisms public-private partnership has a high degree of economic risk.

In this connection, preference is given to the first options of solution of the problem, according to which provides for the implementation of measures aimed at the progressive reduction to the lowest acceptable level of risk the negative impact of economic activities on the environment and the population.

Mechanism for Implementing Environmental Strategy

One of the most important conditions for effective implementation of the strategic developments in the region is to strengthen state regulation in the field of environmental protection. Of particular relevance in this connection acquires the solution of problems of the formation of the economic mechanism of environmental regulation aimed at promoting environmental management and environmental protection, promotion and support of environmentally responsible business, improving the organizational structure of management and legal foundations of environmental protection. From the quality of an appropriate mechanism depends largely the effectiveness of the implementation of medium-and long-term forecasting documents in the field of ecology.

Mechanism for the implementation of the regional program requires a definite complex of legal, economic, organizational, informational, and other measures, which are an integral part of the national environmental policy. Achieving favorable ecological situation as a prerequisite for a decent quality of life and health is possible only on condition of concerted action of regional authorities, business and the public in the field of environmental protection.

Conditions of implementation of environmental strategies should include, in our view, the following elements (Figure 15):

1. Economic mechanism of implementation of environmental policies;
2. Legal support of environmental policy;
3. Environmental financing;
4. Innovative aspects of environmental safety of economic development;
5. The organization of management, coordination and monitoring of the planned environmental protection measures.

Analysis of each of these directions of formation mechanism of the implementation of the environmental strategy requires an independent and sufficiently detailed exposition, therefore, restrict ourselves to brief remarks.

Figure 15. Possible structure of the mechanism of implementation of the environmental strategy
Source: composition of author

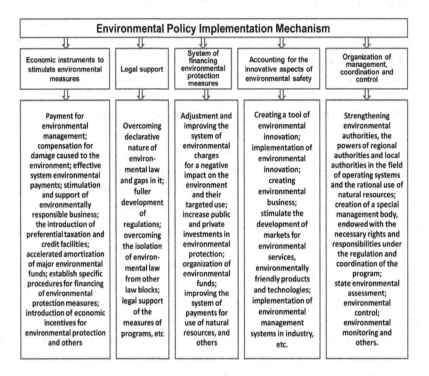

Main directions of the formation mechanism of the implementation of the environmental strategy depend on both the measures taken at the federal level and on the powers of the subject of the Russian Federation, in cooperation with local authorities and, in our opinion, should primarily include:

- Development and implementation of long-term and / or departmental target programs Novosibirsk region in cooperation with environmental problems;
- Development and implementation of projects and plans of the executive authorities of the Novosibirsk region, local governments;
- Implementation of investment projects and plans for development of enterprises that use natural resources;
- Improvement of the legal base in the field of environmental protection and environmental safety;

- Motivation of local governments, enterprises that use natural, scientific and expert community to actively participate in the implementation of planned activities;
- Attracting resources of the federal and local budgets, private sector and civil society organizations for the implementation of programs and projects;
- Maintaining the continuous monitoring and the analysis of the effectiveness of programs and projects.

Implementation of the strategic directions of environmental activities in the Novosibirsk region requires the creation of adequate financing environmental protection, which requires:

- Attracting investment in the environmental sector, mainly due to companies' own funds, increasing the share of equity in natural resources conservation activities;
- A clear delineation between the sources of funding for the protection of ambient between the company's own funds, extra-budgetary and budgetary sources;
- Improving the system of environmental charges and charges for using of natural resources;
- Use of the federal budget, increasing the financing of interregional environmental measures at the expense of the federal budget as co-financing;
- Strengthening the role of regional budgets in financing environmental programs and environmental protection measures, increase in funds for environmental measures as a part the budgets of the subjects of the Federation;
- Improvement of regional environmental funds;
- Increasing investment activity to the resource-saving technologies;
- Attraction of credits of the banks;
- Introduction of obligatory ecological insurance for series of potentially hazardous industries and technologies, etc.

For the purpose facilitating and supplement intraregional environmental financing as key directions are considered the mobilization of domestic resources which are the main source of funding for environmental activities, as well as more effective use of external resources. The main focus should be placed on expanding the budget and resource base and on improving the use of budgetary funds.

CONCLUSION

In general, a large city it is not only big problems (including environmental), but also a great opportunity for their decisions. Almost all environmental problems of the big cities not only have a tendency to be aggravated, but many of them more or less may be solved, as evidenced by the world practice, when the effective solutions for many ecological problems of big cities are being constantly seek and are being find. Formation of smart cities - one of them.

Ensuring sustainable development of the city as one of the ways to eliminate current environmental threats requires greening business activities in various fields, providing for a steady and consistent implementation of systems of technological, administrative, legal and other solutions to improve the efficiency of use of natural resources along with better (or at least preserving) quality of the environment. The success of the greening of the economy also largely depends on the greening of economic management tools and improve their institutional support.

Today's environmental challenges and threats put new challenges to society and require choice of new priorities, based on the principles of "green" economy, according to which economic growth and improving people's well-being must be at the same time reducing negative environmental impact. In other words, economic growth and protection of the environment are mutually reinforcing strategies.

Recommendations were made to improve the management mechanism of the regional eco-economic system. They include development and implementation of long-term and / or departmental target programs Novosibirsk region in cooperation with environmental problems; development and implementation of projects and plans of the executive authorities of Novosibirsk and the Novosibirsk region, local governments; implementation of investment projects and plans for development of enterprises that use natural resources; improvement of the legal base in the field of environmental protection and environmental safety; motivation of local governments, enterprises that use natural, scientific and expert community to actively participate in the implementation of planned activities; attracting resources of the federal and local budgets, private sector and civil society organizations for the implementation of programs and projects; maintaining the continuous monitoring and the analysis of the effectiveness of programs and projects.

Achieving the goals and objectives is aimed at addressing health improvement of the environment within the Novosibirsk region, reducing the anthropogenic load, maintaining acceptable levels of air pollution from stationary and mobile sources of emissions, organization of rational use and protection of water resources, improvement of the treatment of waste production and consumption, as well as identifying possible directions of improving control mechanisms in the field of environmental protection. Orientation to manufacture high-tech products, realization of the project on gasification of industrial and household sectors of the region and other projects provided for in Novosibirsk in the long term, will contribute (along with carrying out the conservation measures) improve environmental performance of the economy of the Novosibirsk region. Exit to the intended target environmental guidelines will testify about the possibility of not only preserving the achieved quality of the environment (primarily due to air quality), but also its progressive improvement.

In accordance with supposed ecological priorities, nature conservation activity in Novosibirsk should be directed at the following: sequential reduction in negative technogenic impact on the environment; prevention of degradation of natural complexes when implementing new investment projects; use of a nature factor for restoring and improving people's health (primarily by development of tourism and creation of recreation zones); increased quality of the drinking water supply; combating underflooding of the territory and feral herd infection (tick-borne encephalitis).

Orientation toward these priorities will make it possible to gradually improve environmental quality in Novosibirsk and, on this basis, implement principles of stable development of the oblast in the mid- and long-term perspective with allowance for solving problems on environmental conservation.

The results of fulfilled research allow to carry out a more informed choice of the main directions of environmental activities in the city under consideration, which avoid the possible risks and the best use of available resources. All of this is a prerequisite for the formulation of priorities in the environmental field and the development of strategies for the protection of the environment as an element of sustainable development of the city as a whole.

REFERENCES

Бурматова, О. П. (2009). Управление воздействием отраслей экономики на окружающую среду: Учебное пособие [Managing the impact of industries on the environment: Textbook]. Novosibirsk: NSU.

Чубик М.П. (2006). *Экология человека: Учебное пособие* [Human Ecology: Textbook]. Tomsk: Publishing House TSU.

Государственный доклад "О состоянии и об охране окружающей среды Новосибирской области в 2015 году." (2016a). [The State Report "On the state and Environmental Protection of the Novosibirsk region in 2015]. Novosibirsk. Retrieved from http://dproos.nso.ru/sites/dproos.nso.ru/wodby_files/files/wiki/2014/12/gosdoklad-2015_0.pdf

Государственный доклад «О состоянии санитарно-эпидемиологического благополучия населения в Новосибирской области в 2015 году». (2016b). [The State Report "On the state sanitary and epidemiological welfare of the population in the Novosibirsk region in 2015"]. Novosibirsk. Retrieved from http://54.rospotrebnadzor.ru/c/document_library/get_file?uuid=ea420274-bb3f-4a95-aef6-620cb759f4fc&groupId=117057

Государственные доклады «О состоянии и об охране окружающей среды Российской Федерации». (n.d.). [State reports "of the Russian Federation on the state and protection of the environment]. Retrieved from http://www.mnr.gov.ru/regulatory/list.php?part=1101

Характеристики основных загрязнителей окружающей среды. (n.d.). [Characteristics of the main pollutants of the environment]. Retrieved from http://www.projects.uniyar.ac.ru/publish/ecostudy/toxic2.html#0

Комбинированное (комплексное) действие ядов. (n.d.). [Combined (complex) action of poisons]. Retrieved from http://www.neonatology.narod.ru/toxicology/kompl_deistvie_jadov.html

Концепция охраны окружающей среды Новосибирской области на период до 2015 года. Утверждена распоряжением Губернатора Новосибирской области от 17.11.2009 N° 283-р. (n.d.). [The concept of environmental protection of the Novosibirsk region for the period up to 2015. Approved by order of the Governor of Novosibirsk region 17.11.2009 number 283-p.]. Retrieved from http://www.nso.ru/sites/test.new.nso.ru/wodby_files/files/migrate/activity/Socio-Economic_Policy/strat_plan /Documents/file895.pdf

Львович Н.К. (2006). *Жизнь в мегаполисе* [Life in the megalopolis]. Nauka.

Лысцов, В.Н., & Скотникова, О.Г. (1991). О возможности взаимного усиления вредных воздействий загрязняющих агентов окружающей среды [On the possibility of a mutually reinforcing harmful effects of polluting environmental agents]. *Журнал Всесоюзного Химического Общества им. Д.И. Менделеева, 1*, 61-65.

Навстречу «зеленой» экономике: пути к устойчивому развитию и искоренению бедности - обобщающий доклад для представителей властных структур. (2011). [Towards to a "green" economy: Pathways to sustainable development and poverty eradication - a synthesis report to representatives of authorities]. UNEP.

Петин, В.Г., & Сынзыныс, Б.И. (1998). *Комбинированное действие факторов окружающей среды на биологические системы* [The combined effect of environmental factors on biological systems]. Obninsk: IATE.

Программа реиндустриализации экономики Новосибирской области до 2025 года (утв. постановлением Правительства Новосибирской области от 01/04/2016 № 89-п). (2016). [Program of reindustrialization of the Novosibirsk Region's economy until 2025 (approved by the Government of the Novosibirsk region 04.01.2016 number 89-p)]. Retrieved from https://www.nso.ru/page/15755

Программа социально-экономического развития Новосибирской области до 2015 г. (n.d.). [The program of social and economic development of Novosibirsk region until 2015]. Retrieved from http://economy.newsib.ru/files/99713.pdf

Сибири, Э. (2009). стратегия и тактика модернизации [Economy of Siberia: The strategy and tactics of modernization]. Novosibirsk: Ankil.

Сонгдо — умный город будущего. (n.d.). [Songdo - smart city of the future]. Retrieved from http://green-agency.ru/songdo-umnyj-gorod-budushhego

Стратегический план устойчивого развития города Новосибирска и комплексные целевые программы. (2004). [Strategic Plan for Sustainable Development of the city of Novosibirsk and complex target programs]. Novosibirsk.

Стратегия социально-экономического развития Новосибирской области до 2025 года. (2007). [Strategy of Socioeconomic development of Novosibirsk oblast until 2025]. Novosibirsk. Retrieved from http://economnso.ru/files/1654.pdf

Умные города будущего: как строить полноценные интеллектуальные города в России. (n.d.). [The Smart Cities of the future: how to build a full-complete smart city in Russia]. Retrieved from http://sk.ru/utility/scripted-file.ashx?GroupKeys=ne t%2F1120292%2F&_cf=callback-standard-detail.vm&_fid=2609466&_ct=page&_ cp=blogs-postlist&_ctt=c6108064af6511ddb074de1a56d89593&_ctc=1549&_ctn =7e987e474b714b01ba29b4336720c446&_cc=0&AppType=Weblog

Умные города: потенциал и перспективы развития в регионах России. Круглый стол (ГУ ВШЭ, Москва). (2014). [Smart Cities: Potential and prospects of development in the Russian regions. Round table. Retrieved from http://issuu.com/ epliseckij/docs/bc9fac678b9405/5?e=7773934/7474790

Урбанизация и ее воздействие на состояние окружающей среды. (2011). [Urbanization and its impact on the environment]. Retrieved from http://phasad. ru/z1.php

В Японии официально открыт "умный город" Фудзисава. (2014). [In Japan, officially opened the "smart city" Fujisawa]. Retrieved from http://hitech.vesti.ru/ news/view/id/6071

Burmatova O.P. (2015a). Nature conservation strategy for regional socioeconomic development. *Region Research of Russia, 5*(3), 286-297.

Burmatova, O. P. (2015b). Environmental and Economic Diagnostics of the Local Production Systems. In Functioning of the local production systems in Bulgaria, Poland and Russia theoretical and economic policy issues. Lodz: Łódź University Press.

Стратегия социально-экономического развития Сибири до 2020 года (Раздел IV. Приоритетные межотраслевые направления развития Сибири - Решение экологических проблем) – Утверждена распоряжением Правительства Российской Федерации от 5 июля 2010 г. № 1120-р. (n.d.). [Strategy of socio-economic development of Siberia until 2020 (Section IV. Priority intersectoral directions of the development of Siberia -. Environmental issues) - Approved by the Federal Government on July 5, 2010 № 1120-p]. Retrieved from http://www. sibfo.ru/strategia/strdoc.php

Hollands R. G. (2008). Will the Real Smart City Please Stand Up? *City, 12*(3), 303-320.

Kuleshov, V.V., & Seliverstov, V.E. (2016). Program for reindustrialization of the Novosibirsk Oblast Economy: Development ideology and main directions of implementation. *Regional Research of Russia, 6*(3), 214-226.

Matugina, E. G., Egorova, A.Y., & Palamarchuk, A. V. (2015, May). To the question of forming and development of ecological entrepreneurship. *SWorld Journal. Economy*, 66-68.

Resolution of the UN General Assembly adopted on 25 September 2015. (n.d.). Retrieved from https://documents-dds-ny.un.org/doc/UNDOC/GEN/N15/291/92/PDF/N1529192.pdf?OpenElement

Seliverstov, V.E. (2012). Regional monitoring: Information management framework for regional policy and strategic planning. *Regional Research of Russia, 2*(1), 60-73.

Seliverstov, V.E., & Melnikova, L.V. (2013). Analysis of Strategic Planning in Regions of the Siberian Federal District. *Regional Research of Russia, 3*(1), 96-102.

ENDNOTES

[1]   Novosibirsk is administrative center of the Siberian Federal District. It has the population of 1,6 million people (for 2016). Novosibirsk is the third city in Russia in terms of population and occupied territories. The leading industries are energy, gas supply, water supply, metallurgy, metalworking, mechanical engineering, they accounted for 94% of total industrial production of the city.

[2]   In this program, a key role belongs to Novosibirsk as a regional center and the city, concentrated basic economic potential of the region.

[3]   Roshydromet - Russian Federal Service for Hydrometeorology and Environmental Monitoring. Web site of Roshydromet: is http://www.meteorf.ru.

[4]   PAP - potential of air pollution - a combination of meteorological conditions that determine the dispersion (accumulation) of impurities coming in the form of emissions from industries and motor vehicles.

[5]   http://greenologia.ru/eko-problemy/goroda/novosibirsk.html - Экологическая ситуация в Новосибирске (The environmental situation in Novosibirsk).

[6]   https://dproos.nso.ru/ - website of the Department of Natural Resources and Environmental Protection of the Novosibirsk region.

[7]   https://dproos.nso.ru/ - website of the Department of Natural Resources and Environmental Protection of the Novosibirsk region.

Chapter 3

How Internet of Things Is Transforming Project Management

Marisa Analía Sanchez
Universidad Nacional del Sur, Argentina

ABSTRACT

Organizations are experiencing a transformation as a consequence of digital technologies such as social, mobile, big data, cloud computing, and internet of things. The transformation presents challenges at several levels, and project management is not an exception. There are changes in the project environment, the power structures, capabilities, skills, and standard practices, just to name a few. Considering the eventual obsolescence of many project portfolio management practices, the aim of this chapter is to discuss the influence of internet of things in this discipline. The analysis departs from rethinking project management insights and describes the impact of smart and connected products considering many dimensions. Recommendations for each PPM stage are developed, followed by a brief discussion of future research directions.

DOI: 10.4018/978-1-5225-3996-4.ch003

Copyright © 2018, IGI Global. Copying or distributing in print or electronic forms without written permission of IGI Global is prohibited.

INTRODUCTION

The most profound technologies are those that disappear. They weave themselves into the fabric of everyday life until they are indistinguishable from it. (Weiser, 1991)

Organizations are experiencing a transformation as a consequence of digital technologies such as social, mobile, big data, cloud computing, and Internet of Things (IoT). The impact of information technology is important even in sectors that are not information intensive such as the agricultural or mining. The digital transformation presents challenges at several levels, namely in leadership, management of markets, management of processes, in how to integrate technologies to transform the organization, and project management. Organizations that do not adapt to this new context will probably lose markets and will vanish. Weill and Woerner (2015) point out that organizations not only fail to take the opportunities given by the digitization but fail to adapt their business models to reflect the economic characteristics and underlying mechanisms of digitization.

Porter and Heppelmann (2014) indicate that Information Technology (IT) is becoming an integral part of the product itself. Embedded sensors, processor, software, and connectivity in products, coupled with a product cloud in which product data is stored and analyzed and some applications are run, are driven dramatic improvements in product functionality and performance.

The evolution to smart, connected products requires a traditional manufacturer to build what is essentially an internal software company (Porter & Heppelmann, 2015). Building software related characteristics is not usually part of the traditional product engineering process. Slama *et al.* (2015) indicate the IoT involves a clash between two worlds in which those in the machine camp and those in the Internet camp will be required to work together to create products that combine physical products with Internet-based application services. Regarding Project Management (PM), the IoT changes the project environment, the power structures, capabilities, skills and standard practices, just to name a few.

Considering the eventual obsolescence of many Project Portfolio Management practices, the aim of this chapter is to discuss the impact and influence of IoT in this discipline. The work is divided in five sections. First, a background on digital transformation helps to put in context the topic of the chapter and explain the influences affecting organizations. Second, the main focus of the chapter is introduced describing how IoT impacts PPM. Then, taking into account the challenges posed by IoT, a discussion for each PPM stage is developed. Future research directions are briefly described. Finally, a conclusion stressing the main challenges for current leaders closes the chapter.

BACKGROUND

Digital Transformation

The corporate strategy is the input to the creation of initiatives, comprising any number of portfolios of programs and projects, which are the vehicles for executing the organization´s strategy. The nature of IoT projects, immersed in a dynamic environment, characterized by a rapid technological change, and part of a digital ecosystem, require looking at projects in a more strategic way. Hence, it is necessary to give an overview of the influence of information technology from a strategic perspective.

In the past, most information technologies adopted by organizations were a means to lower operational costs and increase productivity. Hence, the broad strategic view was that IT strategy must be aligned with the firm's business strategy (Henderson & Venkatraman, 1993). In the last years, the digital infrastructure of organizations and society has radically changed and both researchers and managers have acknowledged that the role of IT has undergone a transformation (Oestreicher-Singer & Zalmanson, 2013). El Sawy (2003) was one of the first to refer to the IT fusion model in which technology is fused in the business environment.

Porter and Heppelmann (2014) provide an historic view to the analysis of the influence of technology, and describe three "waves" as follows:

- **The First Wave of IT (1960s – 1970s):** Automated individual activities in the value chain. The productivity of activities increased because huge amounts of new data could be captured and analyzed in each activity.
- **The Second Wave (1980s – 1990s):** Given by the rise of Internet with its inexpensive and ubiquitous connectivity. This enabled coordination and integration across individual activities, with outside suppliers, channels, and customers; and across geography.
- **The Third Wave (Now):** Of IT is becoming an integral part of the product itself. Embedded sensors, processor, software, and connectivity in products, coupled with a product cloud in which product data is stored and analyzed and some applications are run, are driven dramatic improvements in product functionality and performance.

Porter and Heppelmann (2014) argue that although the IoT shows a new set of technological opportunities, the rules of competence and competitive advantage remain the same. The authors analyze IoT impact on industry structure and the limits of industry to understand the effects of intelligent and connected products. They also describe some strategic options. In a more recent article, the authors analyze the internal implications, such as how the nature of intelligent and connected products redefine the work in each function (product development, IT, manufacturing, logistics, marketing, sales and post-sales service) (Porter & Heppelmann, 2015).

Loebbecke and Picot (2015) use the term "digitization" to refer to changes of established patterns caused by the digital transformation and complementary innovations in economy and society. Digitization penetrates all areas of life and creates new ways of working, communicating and cooperating. Shirky (as cited in (Loebbecke & Picot, 2015)) mentions that connecting individuals, enterprises, devices and governments enables easier transactions, collaboration and social interaction and results in enormous accessible data sources. The interaction between objects adds a multitude of data sources throughout organizations and society. The focus on connected sensors and appliances defines challenges to data flow management. More connections are required with a broad spectrum of organizations and this depends on an adequate management of new relations with stakeholders such as customers, suppliers and rivals.

The improvements in business models that derive from digitization aim to optimize existing processes to increase global efficiency and services and products quality. Digitization enables and makes easier data collection, communication and control activities and in that way reduces transaction costs. However the standardization and massive adoption of these solutions are not sufficient to get a sustainable competitive advantage (Markus & Loebbecke, 2013).

The MIT Sloan Management Review and Deloitte conducted a research and the authors highlight that strategy, not technology, drives digital transformation. Maturing digital businesses are focused on integrating digital technologies, such as social, mobile, analytics and cloud, in the service of transforming how their businesses work (Kane, Palmer, Phillips, Kiron, & Buckley, 2015). The findings of this research show that organizations where digital has transformed processes, talent engagement and business models have a clear and coherent digital strategy. Also, digitally maturing organizations are more comfortable taking risks than their less digitally mature peers.

More recently, a research on designing digital organizations also emphasizes the importance of developing a business strategy that takes advantage of digital technologies (Ross, Sebastian, & Beath, 2016). The authors distinguish two kinds of strategies: a customer engagement strategy which targets to superior, personalized experiences that engender customer loyalty; and a digitized solutions strategy aimed at information enriched products and services that deliver new value for customers. In addition, the research observes that operational excellence is the minimum requirement for doing business digitally.

Internet of Things

In this technology-rich scenario, real world components interact with cyberspace via sensing, computing and communication elements, thus driving towards what is called the Cyber-Physical World (CPW) (Conti et al., 2012). The CPW is characterized by a large number of mobile devices that the users bring around or that are spread into the environment. This scenario enables the automatic observation and measurement of human behavior and the data captured in this way supports data-driven dynamic adaptation of systems in response to observed user habits, preferences and routines.

Various concepts have been coined and promoted by different scholars and technologist. Olson *et al.* present a collective of concepts that are used in depicting future visions of society as affordable by technology (Olson, Nolin, & Nelhans, 2015). In this section, concepts closer to the purpose of the chapter are included: ubiquitous and pervasive computing, and IoT. Ubiquitous computing refers to a society in which human computer interaction is seamlessly and unnoticeably integrated into everyday life. Weiser (1991) describes this concept based on ideas arising from anthropological studies of work practices. Later, IBM coined the term pervasive computing which has a focus more on technological issues. According to Orwat et al. (2008) pervasive computing is associated with the further spreading of miniaturized mobile or embedded information and communication technologies (as cited in Olson, Nolin, & Nelhans, 2015). IoT was introduced by Kevin Ashton in 1999 (Ashton, 2009). Ashton visualized that a physical world can be connected to the internet via sensors and actuators which are capable of providing real time information and hence benefit our lives. With time, this notion has been considerable broadened (Olson, Nolin, & Nelhans, 2015). Gubbi *et al.* (2013) defined IoT as the interconnection of sensing and actuating devices providing the ability to share information across platforms through a unified framework, developing a common operating picture for enabling innovative applications.

Current application domains of IoT include home and personal (control of home equipment, smart sensors in healthcare), enterprise (RFID in retail, industrial ecosystems), utilities (smart grid and household metering), transport (smart traffic, automatic driven vehicles, and intelligent logistics).

Role of Platforms

While considering IoT, it is unavoidable to discuss the role of platforms. Information technology (Internet, mobile, cloud computing, social) has reduced the need to own physical infrastructure and assets giving raise to platform businesses such as Amazon, Uber, Airbnb or eBay. A platform provides the infrastructure and rules for a marketplace that brings together producers and consumers. Platforms comprise four types of players enabling value-creating interactions. The owners of the platforms control their intellectual property and governance (Google owns Android). Providers serve as the platforms' interface with users (mobile devices are providers on Android). Producers create their offerings (for example, apps on Android), and consumers use those offerings (Van Alstyne, Parker, & Choudary, 2016). This shift from pipeline business models to platforms has profound consequences in enterprises. Parker, Van Alstyne and Choudary (2016) have extensively discussed the rise of the platform as a business and organizational model. With the advent of IoT, the concept extends to the "Internet of Things as Platform" to handle device connections and allow users to specify the meaning of the device interactions. For example, smart cities (Wang, He, Huang, & Zhang, 2014), energy grids (Tanoto & Setiabudi, 2016), manufacturing (Woo, Jung, Euitack, Lee, Kwon, & Kim, 2016) or healthcare (Ishii, Kimino, Aljehani, Ohe, & Inoue, 2016) as platforms.

MAIN FOCUS OF THE CHAPTER

Classical Project Management is highly institutionalized and strongly supported by "de facto standards" or "best practices", like PMBok (Project Management Institute, Inc., 2008). On the other hand, the Rethinking Project Management (RPM) concept understands project management as a holistic discipline for achieving organizational efficiency, effectiveness and innovation (Jugdev, Thomas, & Delisle, 2001). RPM perspective adopts a broader conceptualization of projects as being multidisciplinary, having multiple purposes, not always pre-defined, but permeable, contestable and open to renegotiation throughout (Winter, Smith, Morris, & Cicmil, 2006). Svejvig and Andersen (2015) present the results of a structured review of the rethinking project management literature. Through the analysis the authors identified global topics and then categorized each contribution within one of six overarching categories: contextualization, social and political aspects, rethinking practice, complexity and uncertainty, actuality of projects and broader conceptualization. This section aims to highlight a perspective on issues related to project management in the context of IoT. The organization of the presentation is based on the categories identified by Svejvig and Andersen. These categories are rather broad and emerged from the analysis of RPM literature whose concerns are more close to the challenges posed by smart and connected products development.

Contextualization

Contextualization describes how projects need to expand beyond the narrow goals of isolated projects and encourage thinking about projects in a broader context by focusing in the management of multiple projects, the organizational strategy and the project environment (Svejvig & Andersen, 2015).

Management of Multiple Projects

The standard documents of project management currently issued by international project management organizations represent an excellent overview of what the management of a single project includes in its application area (Artto & Kujala, 2008). One limitation of this view is that it does not take into account project interdependencies. In addition, for the case of IoT, several firms participate in projects. Then, the management of a project network or of a business network framework may be more adequate.

Organizational and Ecosystem Strategy

RPM view encourages thinking in organizational strategy, IoT projects need to expand to an ecosystem strategy. For the case of organizations attempting to create value from IoT innovations, the role of stakeholders is essential. In order to define a value proposition, the abilities, knowledge and collaboration of different stakeholders is required. Stakeholders may be defined as individuals, groups and organizations that are affected by or can affect a decision or action (Freeman, 1983). Stakeholder theory recognizes that organizations have obligations not only to shareholders but also to other interest groups such as customers, employees, suppliers and the wider community; amongst many others (Asif, Searcy, Zutshi, & Fisscher, 2013). Addressing the needs of internal and external stakeholders has been identified as a key element of successful product development (Majava, Harkonen, & Haapasalo, 2015). The role of partnerships has also been addressed more thoroughly by digital ecosystem research (Iansiti & Levien, 2004, 2004a). The digital ecosystem theory is adequate to analyse the role of IT in organizations that operate in complex networks and has its origins in complexity theory (Stacey, 1995) and organizational ecology (Hannan & Freeman, 1977). Business ecosystems are networks of organizations that are held together through formal contracting and mutual dependency. The entities of a business ecosystem are structured around core firms, whose centrality is established on the basis of control over the dominant technological architecture or brand that structures value in the ecosystem, or other factors such as product characteristics or geography (Teece, 2007). To summarize, organizations are part of a digital ecosystems, and project success depend on the success of the ecosystem strategy.

Products Part of Broader Systems

IoT-enabled products are part of broader systems. They are composed of components and delivered services depend on other systems. This has been the challenge for Systems Engineering field for many years. Software systems are deployed as part of a large set of components and hence software practitioners are comfortable with a systems thinking approach during all the development processes. For example, during design software teams are aware of issues regarding compatibility and safe integration among software components. These issues are also present during implementation, testing, and deployment. There is a body of experience, methods and tools to cope with systems development. And the challenge for a non-IT organization is to learn from that experience.

Social and Political Aspects

The category social and political aspects include literature with a focus on how social and political processes shape projects, *e.g.* power structures, emotionality and identities (Svejvig & Andersen, 2015). IoT-enabled products offer the opportunity to create new value with data and organizations should have data analytics capabilities. Organizations require expertise at working with large volume of data and also capability of helping leaders reformulate their challenges in ways that big data can tackle. It is also necessary to increase cooperation between the people who understand the problems and the ones who master problem-solving techniques and can generate value from data. Successful information technology organizations will train and recruit people with a new set of skills who can integrate analytic capabilities into their production environments (Davenport, Barth, & Bean, 2012). On the other hand, a data-driven decision making poses managerial challenges related with how decisions are made and who gets to make them (Klotz, 2016). Regarding leadership, Bennis (2013) argues that information-driven transparency will change the way that power is delivered by top leaders, and that leaders need to embrace that transparency.

Another implication of the digital ecosystem environment is that companies can co-create business with people. Co-creation refers to the scenario in which individuals or consumer communities produce marketable value in voluntary activities conducted independently of any established organization, although they may be using platforms provided by such organizations (Karhu, Botero, Vihavainen, Tang, & Hämäläinen, 2011). The management of such platforms needs to address social issues.

Rethinking Practice

Rethinking practice encompasses literature that suggests alternative methods, perspective and approaches to rethink the ways in which practitioners work with projects. Regarding IoT, there are several issues that emerge and have implications on project management: the shift to a product-as-a-service business model; the focus on customer success; and data analytics as a source of value creation.

Product-As-A-Service Business Model

Companies in search of growth are working on integrating products and services. Reim *et al.* (2015) provides a literature review to understand implementation of product-service systems business models and tactical practices. IoT-enabled products which can track customer usage and satisfaction represents an opportunity for manufacturers to leverage technology to create new business models that shift focus from standalone products to service-based offerings. As described by Porter and Heppelman (2015) the new data available to companies, together with new configurations and capabilities of smart, connected products, is restructuring the traditional functions of business and this transformation started with product development.

Focus on Customer Success

Users are at the center of intelligent environments, in that respect this area overlap with the efforts of the scientific community focused on Person-Centric Computing (Augusto, Callaghan, Cook, Kameas, & Satoh, 2013). The introduction of technology has to be sensitive to the user and abide to the principle that the human is the master and the computer the slave and no the other way round (Dertouzos, 2001) cited in (Augusto, Callaghan, Cook, Kameas, & Satoh, 2013)). Smart, connected products capture usage data that allows segmenting customers, customizing products, setting prices to better capture value, and extending value-added services (Porter & Heppelmann, 2014). Organizations that focus on customer success and experience are more prepared to enhance differentiation.

Analytics as a Competitive Advantage

Project managers should integrate data analytics as a core capability since it supports many activities from design, production and even innovation. The creation of new value with data requires data capabilities. Social networks, mobile and IoT products generate huge amount of data that may be used to improve product or services design and support a differentiation strategy. Hence, organizations must add talent in data analytics area. Software companies providing e-commerce solutions have long understand the power of data analytics in generating customer value (Porter & Heppelmann, 2015). For the case of Big Data, Davenport (2012) describes several ways it can be used to enhance a business, such as making routine business decisions faster, supporting new decisions and developing new product and services. Any of these applications may contribute to cost savings or improve differentiation. Progress in data storage and processing technology led to more data-driven decision making. Brynjolfsson *et al.* (2016) found that between 2005 and 2010, the share of manufacturing plants that adopted data-driven decision-making early tripled to 30 percent but the rapid diffusion is uneven. McAfee and Brynjolfsson describe five management challenges related with Big Data that are related with leadership, talent management, technology, decision making and culture (McAfee & Brynjolfsson, 2012).

Complexity and Uncertainty

The category complexity and uncertainty consists of contributions that deal with the increasing uncertainty and complexity in projects and project environments (Svejvig & Andersen, 2015). Shenhar and Dvir (2007) have proposed a framework known as the "Diamond of Innovation" which provides a framework for project classification. Each one of its dimensions of novelty, technology, complexity and pace consists of four possible project categories, and by selecting a category in each dimension, one creates a specific diamond-shaped view for each project. Once a classification is selected, the model helps identify the unique impact of each dimension, and provides recommendations for a preferred style of management for increasing the likelihood of the project's success. Novelty refers to how new is the product to the market, users, and customers. Technology innovation describes how much new technology is used and it impacts product design, development, testing, and the required technical skill. Complexity represents the complexity of the product or the organization and impacts the degree of formality and coordination needed to effectively manage the project. Pace denotes the urgency of the innovation and impacts the time management and autonomy of the project management team. Table 1 summarizes the values that each dimension may assume. The dimensions proposed by the framework take into account not only the organizational readiness to develop the project but also the external issues related to suppliers, subcontractors or the market. Then, it results in an adequate framework to support the analysis of projects developed in an ecosystem environment. In what follows, there is a discussion about complexity and uncertainty based on the "Diamond of Innovation" dimensions.

Table 1. Dimensions of the Diamond of Innovation Model

Dimension	Categories in Each Dimension
Complexity	Assembly: subsystem, performing a single function. System: collection of subsystems, multiple functions. Array: widely dispersed collection of systems with common mission.
Technology	Low-tech: no new technology is used. Medium-tech: some new technology. High-tech: all or mostly new, but existing technologies. Super high-tech: necessary technologies do not exist at project initiation.
Pace	Regular: delay not critical. Fast-competitive: time to market is important for the business. Time-critical: competition time is crucial for success-window of opportunity. Blitz: crisis project-immediate solution is necessary.
Novelty	Derivative: improvement of an existing product. Platform: a new generation of existing product line. Breakthrough: a new-to-the world product.

Source: Adapted from (Shenhar & Dvir, 2007)

Novelty of Innovation

Even digital technologies pose some opportunities, there are uncertainties about the market needs and hence which is the right value proposition to develop. There are no clear rules about how to proceed. Slama *et a*l. (2015) suggest building a minimum viable product and get to market fast. Hence, firms have first-mover advantages by collecting and accumulating product data and using it to improve product and services. Companies that aim to produce smart products require getting new products to market quickly and responding fast to customer need. A first-mover strategy may imply shorter development cycles and the requirement to optimize development. And this drives to the next dimension of the diamond, the pace of innovation.

Pace of Innovation

The pace measurement captures the urgency of time in the success of the project (Shenhar, Holzman, Melaned, & Zhao, 2016). The pace of digital improvement is making strategy complex. In general, IoT-enabled solutions are time-critical since the failure to meet the project milestone deadlines can result in project failure because a window to opportunity will close. For many years, the software industry had evolved software development methods from the traditional waterfall model to agile product development process. The software industry was motivated not only by the need to shorten development cycles but to develop solutions that satisfy the customer requirements. Agile product-development processes emphasize daily collaboration between developers and marketers, weekly delivery of enhancements, continual course corrections, and ongoing testing of customer satisfaction. While this is quite incorporated in any software team, IoT's project team are integrated by people from different backgrounds (engineers, software developers) and cultures so the adoption of agile techniques may not necessarily occurs in the short term. Agile requires training, behavioral change and is not adequate for all innovation types or organizations (Conforto, Salum, Amaral, da Silva, & Magnanini de Almeida, 2014). For example, public agencies and governmental organizations have been slow in adopting agile practices (Nuottila, Aaltonen, & Kujala, 2016). Also, agile methods are not recommended for system critical projects such as those related with nuclear facilities and power grids.

Technology Innovation

Most IoT developments involve new technology (*e.g.* smart homes) and even critical technologies that do not exist (*e.g.* energy grids).

Complexity

The amount of complexity is given by how complex is the product. Smart and connected products may range from discrete components (for example, a thermostat) or an array of widely dispersed collection of systems (for example, smart grid).

The Actuality of Projects

the actuality of projects covers literature that underlines the need for empirical studies of projects as its own point of departure (Svejvig & Andersen, 2015). To review this topic a retrieval of literature related with project management and IoT is analyzed.

The review aims to analyze contributions that have emerged as a result of challenges posed by IoT projects. For this reason, the key terms are based on Svejvig and Andersen (2015) review of Rethinking Project Management literature with the addition of the theme "internet of things". Considering the novelty of these topics, the search strings include "project management" and "internet of things" (Sevjvig and Andersen do not consider the term "project management" as a unique term). Also, given that some proposals may be included in digital transformation literature; both "internet of things" and "digital transformation" search strings were considered. Tables 2, 3 and 4 depict a summary of the publications search in Scopus for the last five years (2012-May 2017). The search strings included in Table 5 provided zero results.

Papers are categorized according to their type (descriptive, prescriptive, conceptual, theory, other) (Ahlemann, El Arbi, Kaiser, & Heck, 2013). Descriptive research answers the questions regarding what and how as well as Yes/No questions. Prescriptive research seeks to help people solve practical problems by developing and testing artifacts (method, model, framework, ontology, reference model, system). Conceptual papers present assumptions, premises, axioms, assertions without empirical work. A theory should fulfill three criteria: (1) a theory must have clear constructs; (2) the relationships between the constructs must be defined; and (3) a theory must be testable. Finally, a category referred to as Other includes Literature analysis, editorials, reports, book reviews, and calls for papers/abstracts/participation. Table 6 depicts the result of the categorization. Only 7 contributions were classified as descriptive (empirical).

Broader Conceptualization

Broader conceptualization deals with contributions that offer alternative perspectives on, for example, projects, project management and project success, outline how the field is broadening beyond is current limits or describe the existing perspectives within the field. Prescriptive, conceptual and review categories in Table 6 may offer a broader conceptualization (36 papers). These papers refer to specific issues but none of them cover the impact of IoT in PM.

SOLUTIONS AND RECOMMENDATIONS

In this section, an overview of each Portfolio Project Management phase is provided with a discussion about IoT impact. The framework proposed by Bible and Bivins (2011) is used because it takes into account a strategic and portfolio management perspective. In addition, an operations and maintenance phase is considered. The

Table 2. Publications by year

Search expression (Scopus search format)	2012	2013	2014	2015	2016	2017	Total
TITLE-ABS-KEY ("digital transformation") AND TITLE-ABS-KEY ("project management")	-	-	-	-	3	-	3
TITLE-ABS-KEY ("internet of things") AND TITLE-ABS-KEY ("project management")	1	4	6	9	14	2	36
TITLE-ABS-KEY ("internet of things") AND TITLE-ABS-KEY ("conventional project management")	-	-	-	1	-	-	1

Table 3. Publications by subject areas. Scopus classifies each publication in one or more subject areas

Search expression (Scopus search format)	Business, Management and Accounting	Computer Science	Engineering	Mathematics	Social Sciences	Decision Sciences	Earth and Planetary Sciences	Energy	Pharmacology, Toxicology and Pharmaceutics
TITLE-ABS-KEY ("digital transformation") AND TITLE-ABS-KEY ("project management")	1	1	1	1	1	-	-	-	-
TITLE-ABS-KEY ("internet of things") AND TITLE-ABS-KEY ("project management")	-	23	11	10	-	4	2	1	1
TITLE-ABS-KEY ("internet of things") AND TITLE-ABS-KEY ("conventional project management")	-	1	-	1	-	-	-	-	-

Table 4. Publications by type (Scopus classification)

Search expression (Scopus search format)	Article	Conf. review	Conf. paper	Review	Total
TITLE-ABS-KEY ("digital transformation") AND TITLE-ABS-KEY ("project management")	2	1	-	-	3
TITLE-ABS-KEY ("internet of things") AND TITLE-ABS-KEY ("project management")	9	8	17	2	36
TITLE-ABS-KEY ("internet of things") AND TITLE-ABS-KEY ("conventional project management")	-	-	1	-	1

assumption underlying the portfolio management project is that products are released and after-sales support requires fewer resources than those required during development. That is, the need of resources tends to decline in traditional projects. However, IoT-enabled products may require additional resources after deployment. Therefore, a solution operation and maintenance phase is included.

As mentioned earlier in this chapter, the nature of IoT projects, immersed in a dynamic environment, characterized by a rapid technological change, and part of

Table 5. Search strings (Scopus format) separated by ";" that produced zero results

TITLE-ABS-KEY ("digital transformation") AND TITLE-ABS-KEY ("rethinking project")) ; TITLE-ABS-KEY ("digital transformation") AND TITLE-ABS-KEY ("complexity theory") AND TITLE-ABS-KEY ("project management") ; TITLE-ABS-KEY ("internet of things") AND TITLE-ABS-KEY ("complexity theory") AND TITLE-ABS-KEY ("project management"); TITLE-ABS-KEY ("digital transformation") AND TITLE-ABS-KEY ("reinventing project") ; TITLE-ABS-KEY ("digital transformation") AND TITLE-ABS-KEY ("beyond project"); TITLE-ABS-KEY ("digital transformation") AND TITLE-ABS-KEY ("conventional project management") ; TITLE-ABS-KEY ("digital transformation") AND TITLE-ABS-KEY ("project management theory") ; TITLE-ABS-KEY ("digital transformation") AND TITLE-ABS-KEY ("perspectives on projects"); TITLE-ABS-KEY ("digital transformation") AND TITLE-ABS-KEY ("project management") AND TITLE-ABS-KEY ("complexity theory"); TITLE-ABS-KEY ("internet of things") AND TITLE-ABS-KEY ("project management") AND TITLE-ABS-KEY ("complexity theory") ; TITLE-ABS-KEY ("internet of things") AND TITLE-ABS-KEY ("rethinking project"); TITLE-ABS-KEY ("internet of things") AND TITLE-ABS-KEY ("project management") AND TITLE-ABS-KEY ("complexity theory"); TITLE-ABS-KEY ("internet of things") AND TITLE-ABS-KEY ("reinventing project"); TITLE-ABS-KEY ("internet of things") AND TITLE-ABS-KEY ("project management theory"); TITLE-ABS-KEY ("internet of things") AND TITLE-ABS-KEY ("perspectives on projects"

Table 6. Type of research contribution

Search expression (Scopus search format)	Descriptive (empirical)	Prescriptive	Conceptual	Review	Conf. Proc.
TITLE-ABS-KEY ("digital transformation") AND TITLE-ABS-KEY ("project management")	1	1	-	-	1
TITLE-ABS-KEY ("internet of things") AND TITLE-ABS-KEY ("project management")	6	15	5	2	8
TITLE-ABS-KEY ("internet of things") AND TITLE-ABS-KEY ("conventional project management")	-	1	-	-	-

a digital ecosystem, require looking at projects in a more strategic way. In order to support this vision, Shenhar and Dvir's "Diamond of Innovation" (Shenhar & Dvir, 2007) appears as necessary framework to classify a project and recommend a managerial style. The classification depends on the project and the organization. A project may represent a high technology level for an organization but a low level for another. For example, Shenhar and Dvir (2016) describe that the technology used in Boeing's Deamliner project represented a "next generation in an existing line of projects" for Boeing but for subcontractors, it was new. Hence, it is not possible to define a generic innovation diamond. However, for the sake of illustrative purposes, Table 7 briefly describes four projects and Figure 1 depicts the innovation diamonds for each case. The classification of projects early at this stage provides clues to a recommended style of management, activities and resources required by the project.

Table 7. Brief description of illustrative projects

Smart learning thermostat. A connected thermostat that can be controlled through an Internet connection and is able to learn when each room of a house is normally occupied and automatically schedule the heating for that room appropriately.
Smart parking solution for smart cities to allow citizens to find available parking spots. Assume the solution should be available promptly.
Smart metering project aims to provide accurate, near real-time consumption information to end consumers; the potential for users to save money and reduce emissions using consumption management tools and more highly differentiated tariffs. The project is owned by the government and it is immersed in a complex ecosystem including multiple smart meter manufacturers, communication service providers, data service providers, electricity suppliers, and meter operators.
Cross-energy management project aims to build the IT architecture for energy supply optimizing multiple heterogeneous energy resources and consumers. The ecosystem is integrated by company and private consumers, electricity suppliers and grid operators.

Figure 1. Projects classification based on the Diamond of Innovation Model

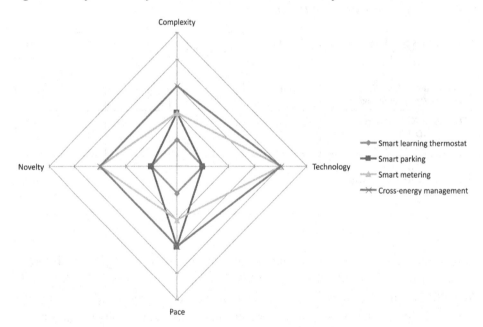

Strategic Phase

During the strategic phase the organization establishes or revises its strategic plan, including mission, vision, goals, and objectives. This phase provides the foundation for effective PPM as the strategic plan establishes the goals and objectives to which project portfolios contribute. The organization should define the strategic importance of connected products in the context of the strategy. The IoT may contribute to define new business or to improve internal process. Also, at this stage CEOs should work on partnerships.

According to Ron Adner (Klotz, 2016), given the relevance of partnerships in a IoT development, the CEO-level should establish partner relationships for the project to be feasible. The answer to this question depends on to the extent that the project depends on collaborating in a new ways with other partners. If the answer is very much then management of co-innovations is an issue (Klotz, 2016). The transition to a successful digital enterprise requires expertise in using digital technologies such as cloud, social, mobile and analytics. In the case of smart products building software related characteristics is not usually part of the traditional product engineering process.

The partnerships concept may go further when the organization becomes part of an ecosystem. In that case, CEOs should define an ecosystem strategy that takes into account expectations, requirements, and capabilities of all the partners across the ecosystem. For each partner there should be a definition of what connected things will develop, and which information and services will deliver.

Screening Phase

The screening phase begins the process of soliciting proposals for potential projects with a reasonable expectation of contributing to goals and objectives. Potential projects that support strategic goals and objectives and seem to be feasible are included in the list of candidate projects that will be evaluated during the selection phase (Bible & Bivins, 2011). In a portfolio approach to project management, projects that are already in progress will be assessed as well to determine whether they continue or should be postponed or cancelled. Also, mandatory projects that arise as a government requirement are included in the pool of candidate projects. The screening phase provides the opportunity to create and analyze the business case (also known as project proposal) for each project against the criteria and assumptions.

Given the requirement of shorter development cycles, this poses restrictions to all project development phases. In particular, during selection phase when projects are prioritized, managers should derive a prioritized list of projects. For the case of IoT products, the window opportunity is shorter and hence they might start as soon as possible.

Screening criteria represent factors against which the pool of potential projects is evaluated to determine whether a business case is sufficient to include the project in the candidate project pool. Screening criteria and assumptions provide a scenario under which the project portfolio will be selected.

Each organization may choose the most convenient tools to develop a project specification according to the organizational available tools, expertise or preferences. However, the Business Model Canvas proposed by Osterwalder and Pigneour (2010) has proven its effectiveness on the preliminary definition of projects. It is a tool to describe the way value is created and delivered in an organization. Consisting in nine building blocks arranged in a template, the model offers a visual representation that allows a better comprehension of business and its dynamics. In the center of the template there is the value proposition which describes the products or services provided by the company to satisfy the needs of one or several customer segments. By designing different channels, the company reaches and communicates with its customers, delivering value. Customer relationships describe particular relationships the company wants to establish with their customers and the means to do so. The complete interaction between these four blocks results in revenue streams which take into consideration the cash that customers generate when paying for the value that is being offered.

The Innovation Canvas (Ahmed, et al., 2014) is a framework that guides teams to develop integrated product designs and business models. There are some different templates to guide the value proposition development (Canvanizer 2.0, 2017; Willness & Bruni-Bossio, 2017). The canvas itself supports a number of design tools, methods, and approaches (Kline et al., 2013). The Innovation Canvas incorporates all the themes of the business Model Canvas in the Market quadrant. The Explore, Ideate and Design quadrants have been added to include the key themes of a meta-model of a system and are described as follows:

- **The Explore Section:** Provides a detailed description of the project goals, collect stakeholder and voice of the customer input, clarify benefits and values of a product and provide context for the project.
- **The Ideate Section:** Describes what the product will do and the external systems that the product interacts with.
- **The Design Section:** Includes a formal requirements analysis, key components or modules, critical success factors of the product, and critical risks of product failure that require special attention.
- **The Market Section:** Includes the nine constructs from the Business Model Canvas (Osterwalder & Pigneur, 2010).

Selection Phase

Each project contributes to the realization of different goals, and some projects may contribute better than others. The objective of the selection phase is to derive a portfolio of projects providing maximum benefit subject to resource constraints and other limitations imposed by the organization. The challenges in PPM selection are how to assess contribution to strategic goals, the one-time portfolio implementation costs (infrastructure, labor, services), the future operational (and maintenance) expenditures entailed (the costs that the organization will incur after closing a project), and environmental and social impacts. In what follows, there is a brief description of how different methods tackle these issues.

There are many possible methodologies that can be used in selecting a portfolio and evaluating projects. A detailed description of methods is out of the scope of this chapter. Methodologies range from easy to use ones that require little information but fail to assess a portfolio of projects, to more sophisticated techniques which limitation may be amount of data required to generate a prioritized list of projects.

Whatever the business case definition approach used by an organization, there are some challenges that arise for IoT products. In order to define the cost structure it is necessary to calculate costs across all parties in the network. Hence, access to information and confidence among partners is required in order to calculate accurate costs. In addition, IoT-enabled connected products require a backend infrastructure that generates operating costs on an ongoing basis (Slama, Puhlmann, Morrish, & Bhatnagar, 2015).

Implementation Phase

The implementation phase transitions the approved project portfolio into execution. Projects are monitored in order to track their development (updating costs and benefits estimates to detect deviations) and re-prioritize them when strategic goals (or their target values) change or new initiatives appear.

As mentioned earlier in the chapter, smart products are dominated by shorter development cycles. Sting *et al.* (2015) address the problem of accelerating product cycles by prioritizing project delivery. The authors mention two characteristics that make optimal delivery times elusive: uncertainty (for example, they develop a new product function whose feasibility has not been established); and the workers have information about the status of their project tasks that is not observable to anyone but themselves, which many don´t share. The authors describe a management innovation at Roto Frank, a German company that produces hardware for industrial and residential window and doors. While this is not a case of IoT-enabled products the difficulties of project planning and execution referred to by authors are valid for smart product development. The management innovation is based on establishing psychological safety and encouraging cooperative behavior by emphasizing interdependence among workers. The control chart was relocated from the meeting room to each project desk where the engineer's critical tasks for every day of the week would be clearly written. In addition, engineers were asked to put up a red flag whenever a critical task was becoming late enough to affect other tasks. Management promised that anyone who raised a red card would not be criticized, and would receive help within 30 minutes. This innovation enabled the organization to react more rapidly to problems.

Evaluation Phase

Projects are monitored in order to track their development (updating costs and benefit estimates to detect deviations) and re-prioritize them when strategic goals (or their target values) change, new initiatives appear, or projects are finished. The purpose of the evaluation phase is to re-prioritize projects and decide whether to cancel, postpone, delay or add projects.

Project managers measure project performance through the triple constraints of scope, cost, and schedule and make adjustments if performance is outside of tolerances. At the same time, the organizational strategy is in constant update as new goals are set or existing goals are dismissed as a result of changes in organizational strategy. Hence, some projects may become obsolete; or newly defined goals may not be supported by any project. Also, newly defined goals may not be supported by any project. Both of these situations reveal that some projects are not creating business value and provide the basis for determining modifications during the evaluation phase. While project managers report on the individual performance of projects, portfolio managers, analyze the overall performance to determine areas of concern and gather information to support decision taking during the evaluation phase.

Operations and Maintenance Phase

The Close Project Phase, as described in the PMIBook is the process of finalizing all activities across all of the Project Management Process Groups to formally complete the project. The key benefit of this process is that it provides lessons learned, the formal ending of project work, and the release of organization resources to pursue new endeavors (Project Management Institute, Inc., 2008). While this process is necessary, smart products lifecycle is different and requires more resources after deployment. In the context of a software development, maintenance is required to fix defects, adapt to changes in the environment, and meet new or changed user requirements. Software cost estimation techniques are based on source lines of code as the software size input measure, due to high correlation with effort; and function points as a measure of software size by functions or modifications to functions. On the other hand, IoT relevant activities are operating the IoT platform; collecting data from the product and end-user behaviour in real time; analyze large volumes of data, and share it with other stakeholders. Costs related with these activities are not adequately estimated using software approaches. More research is required to answer the question of which are appropriate approaches and techniques to support cost calculations.

FUTURE RESEARCH DIRECTIONS

As mentioned previously in this chapter, strategy, not technology, drives digital transformation (Kane, Palmer, Phillips, Kiron, & Buckley, 2015). Also, forecasts suggest that technology will evolve, software will be cheaper an even free (*e.g.* open-source solutions). Therefore, the challenges seem to lay in having the capabilities to get value from technology. In what follows, there is a list of firm specific capabilities that can be a source of advantage, and studies that will expand both the scientific and practice-oriented foundations of these themes are encouraged:

- **Leadership:** The digital transformation context requires new ways of leading. Companies need leaders who can spot a great opportunity, understand how a market is developing, think creatively and propose truly novel offerings, articulate a compelling vision, persuade people to embrace it and work hard to realize it, and deal effectively with all stake holders (Davenport, Barth, & Bean, 2012). Research on new models for understanding and practicing leadership in a complex world is suggested.
- **Collaborative Culture:** IoT developments are knowledge intensive projects, sometimes agile methodologies are used, or they involve co-production. In any of these scenarios, collaboration is a critical success factor. As summarized by Lloyd-Walker *et at.* (2014), relationships are generally enhanced and reinforced through collaborative problem-solving that increases absorptive capacity and generates new knowledge about the project context. This begs the question on how to foster collaboration among groups.
- **Partnerships:** Can be a source for new organizational learning and can enable firms to bring new strategic assets into the firm from external sources. Closely related to collaborative projects, partnership focus on external actors. Gama *et al.* (2017) considers inter-organizational technology development is

problematic because firms lack sufficient partner understanding and struggle with aligning their project management practices with those of their partners. The authors identify PM practices of coordination and control to fit the contingencies of each type of partner collaboration and set the stage for future research.

- **Management of a Project Network and the Management of a Business Network:** As discussed earlier in this chapter, these approaches would help to think about projects in a broader context. Examples of themes from existing literature are found in (Artto & Kujala, 2008).
- **Data Analytics:** One of the most critical resources today is the availability of qualified data scientists capable to apply statistical and data mining techniques, cooperate with the people who understand the problems and create new value from data. Data flowing to and from products provides a picture of product use, which features customers prefer and identify the reason for failure. PM should adapt to integrate these insights in their practices. Research on adequate team organization, how decisions are taken.

Figure 2. Number of articles retrieved in a Scopus search using the term "Internet of Things". Contributions are classified according to the subject area of interest (Hard and Social sciences), and the ratio series illustrates the relationship between the number of articles in each area

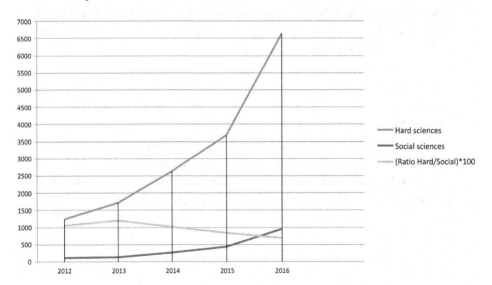

CONCLUSION

The rapid diffusion of IoT technology is reflected by the offering of new services, new opportunities for innovation based on usage and customer behavior data, and the emergence of disruptive business models even before any academic conceptualization. At the same time, PM has long been belatedly in adapting to a broader context. The digital transformation accentuates the need to update best practices. The literature review provides a small number of contributions on the intersection of project management and IoT (in the academic world). On the other hand, there is an increasing interest in IoT within social sciences subject area. Figure 2 shows the number of retrieves articles in a Scopus search using the term "Internet of Things", including the period from 2012 to 2016. The series "Hard sciences" groups contributions in engineering and computer science areas; the series "Social sciences" includes papers in social sciences, business, management and economy. The trend of the ratio over time shows IoT topic is becoming of interest in both research areas. This observation is consistent with the fact that a common theme along the chapter was the increasing relevance of managerial and cultural aspects. This is not surprising since technology evolves much faster than an organization ability to adapt to it. Leaders need to reframe the way strategy is formulated and define the key steps to leverage the entire organization behind a well-defined vision.

REFERENCES

Ahlemann, F., El Arbi, F., Kaiser, M., & Heck, A. (2013). A process framework for theoretically grounded prescriptive research in the project management field. *International Journal of Project Management, 31*(1), 43–56. doi:10.1016/j.ijproman.2012.03.008

Ahmed, J., Rogge, R., Kline, W., Bunch, R., Mason, T., & Wollowski, M. (2014). The innovation canvas: An instructor's guide. In *121st ASEE Annual Conference and Exposition* (pp. 1-12). Indianapolis, IN: American Society for Engineering Education.

Artto, K., & Kujala, J. (2008). Project business as a reseach field. *International Journal of Managing Projects in Busines, 1*(4), 469–497. doi:10.1108/17538370810906219

Ashton, K. (2009). That "Internet of Things" thing. *RFID Journal, 22*, 97–114.

Asif, M., Searcy, C., Zutshi, A., & Fisscher, O. (2013). An integrated management systems approach to corporate social responsibility. *Journal of Cleaner Production, 56*, 7–17. doi:10.1016/j.jclepro.2011.10.034

Augusto, J., Callaghan, V., Cook, D., Kameas, A., & Satoh, I. (2013). Intellignet Environments: a manifesto. *Human-centric Computing and Information Sciences, 3*(12).

Bennis, W. (2013). Leadership in a Digital World: Embracing Transparency and Adaptive Capacity. *Management Information Systems Quarterly, 37*(2), 635–636.

Bible, M., & Bivins, S. (2011). *Mastering Project Portfolio Management. A Systems approach to Achieving Strategic Objectives.* Fort Lauderdale, FL: J. Ross Publishing.

Brynjolfsson, E., & Mcelheran, K. (2016). The Rapid adoption of Data-Driven Decision Making. *The American Economic Review, 106*(5), 133–139. doi:10.1257/aer.p20161016

Canvanizer 2.0. (2017). *Canvanizer.* Retrieved from https://canvanizer.com/new/open-innovation-canvas

Conforto, E., Salum, F., Amaral, D., da Silva, S., & Magnanini de Almeida, L. (2014). Can Agile Project Management be Adopted by Industries Other than Software Development? *Project Management Journal, 45*(3), 21–34. doi:10.1002/pmj.21410

Conti, M., Das, S., Bisdikian, M., Kumar, M., Ni, L., Passarella, A., ... Zambonelli, F. (2012). Looking ahead in pervasive computing: Challenges and opportunities in the era of cyberphysical convergence. *Pervasive and Mobile Computing*, *8*(1), 2–21. doi:10.1016/j.pmcj.2011.10.001

Davenport, T., Barth, P., & Bean, R. (2012). How "Big Data" Is Different. *MIT Sloan Management Review*, 22–24.

Dertouzos, M. (2001). Human-centered Systems. In P. Denning (Ed.), The Invisible Future (pp. 181-192). ACM Press.

El Sawy, O. (2003). The IS Core IX. The 3 Faces of IS Identity: Connection, Immersion, and Fusion. *Communications of the Association for Information Systems*, *12*, 588–598.

Freeman, R., & Reed, D. L. (1983). Stockholders and stakeholders: A new perspective on corporate governance. *California Management Review*, *25*(3), 88–106. doi:10.2307/41165018

Gama, F., Sjödin, D., & Frishammar, J. (2017). Managing interorganizational technology development: Project management practices for market- and science-based partnership. *Creativity and Innovation Management*, *26*(2), 115–127. doi:10.1111/caim.12207

Gubbi, J., Buyya, R., Marusic, S., & Palaniswami, M. (2013). Internet of Things (IoT): A vision, architectural elements, and future directions. *Future Generation Computer Systems*, *29*(7), 1645–1660. doi:10.1016/j.future.2013.01.010

Hannan, M., & Freeman, J. (1977). The population ecology of organizations. *American Journal of Sociology*, *82*(5), 929–964. doi:10.1086/226424

Henderson, J., & Venkatraman, N. (1993). Strategic Alignment: Levering Information Technology for Transforming Organizations. *IBM Systems Journal*, *32*(1), 4–16. doi:10.1147j.382.0472

Iansiti, M., & Levien, R. (2004). *The keystone advantage: what the new dynamics of business ecosytems mean for strategy, innovation and sustainability*. Boston: Harvard BusinessSchool Press.

Iansiti, M., & Levien, R. (2004a). Strategy as ecology. *Harvard Business Review*, *82*, 68–78. PMID:15029791

Ishii, H., Kimino, K., Aljehani, M., Ohe, N., & Inoue, M. (2016). An Early Detection System for Dementia using the M2M/IoT Platform. *Procedia Computer Science*, *96*, 1332–1340. doi:10.1016/j.procs.2016.08.178

Jugdev, K., Thomas, J., & Delisle, C. (2001). Rethinking project management old truths and new insights. *Project Management*, 36-43.

Kane, G., Palmer, D., Phillips, A., Kiron, D., & Buckley, N. (2015). *Strategy, not technology, drives digital transformation*. MIT Sloan Management Review.

Karhu, K., Botero, A., Vihavainen, S., Tang, T., & Hämäläinen, M. (2011). A Digital Ecosystem for Co-Creating Business with People. *Journal of Emerging Technologies in Web Intelligence*, *3*(3), 197–205. doi:10.4304/jetwi.3.3.197-205

Kline, W., Hixson, C., Mason, T., Brackin, P., Bunch, R., & Dee, K. (2013). The Innovation Canvas - A Tool to Develop Integrated Product Designs and Business Models. In *120th ASEE Annual Conference and Exposition* (pp. 1-11). Atlanta, GA: American Society for Engineering Education.

Klotz, F. (2016, August). Navigating the Leadership Challenges of Innovation Ecosystems. *MIT SMR Frontiers*, 1-5.

Lloyd-Walker, B., Mills, A., & Walker, D. (2014). Enabling construction innovation: The roles of a no-blame culture as a collaboration behavioural driver in project alliances. *Construction Management and Economics*, *32*(3), 229–245. doi:10.1080/01446193.2014.892629

Loebbecke, C., & Picot, A. (2015). Reflections on societal and business model transformation arising from digitization and big data analytics: A research agenda. *The Journal of Strategic Information Systems*, *24*(3), 149–157. doi:10.1016/j.jsis.2015.08.002

Majava, J., Harkonen, J., & Haapasalo, H. (2015). The relations between stakeholders and product development drivers: Practitioners' perspectives. *International Journal of Innovation and Learning*, *17*(1), 59–78. doi:10.1504/IJIL.2015.066064

Markus, L., & Loebbecke, C. (2013). Commositized digital processes and business community platforms: New opportunities and challenges for digital business strategies. *Management Information Systems Quarterly*, *37*(2), 649–653.

McAfee, A., & Brynjolfsson, E. (2012). Big Data: The Management Revolution. *Harvard Business Review*, *10*, 60–79. PMID:23074865

Nuottila, J., Aaltonen, K., & Kujala, J. (2016). Challenges of adopting agile methods in a public organization. *International Journal of Information Systems and Project Management*, *4*(3), 65–85.

Oestreicher-Singer, G., & Zalmanson, L. (2013). Content or Community? A Digital Business Strategy for Content Providers in the Social Age. *Management Information Systems Quarterly*, *37*(2), 591–616. doi:10.25300/MISQ/2013/37.2.12

Olson, N., Nolin, J., & Nelhans, G. (2015). Semantic web, ubiquitous computing, or internet of things? A macro-analysis of scholarly publications. *The Journal of Documentation*, *71*(5), 884–916. doi:10.1108/JD-03-2013-0033

Orwat, C., Graefe, A., & Faulwasser, T. (2008). Towards pervasive computing in health care - A literature review. *BMC Medical Informatics and Decision Making*, *8*(26). PMID:18565221

Osterwalder, A., & Pigneur, Y. (2010). *Business Model Generation: A Handbook for Visionaries, Game Changers, and Challengers*. Hoboken, NJ: John Wiley & Sons, Inc.

Parker, G., Van Alstyne, M., & Choudary, S. (2016). *Platform Revolution. How networked markets are transforming the economy and how to make them work for you*. New York: W. W. Norton & Company.

Porter, M., & Heppelmann, J. (2014). How Smart, Connected Products Are Transforming Competition. *Harvard Business Review*, 4–23.

Porter, M., & Heppelmann, J. (2015). How Smart, Connected Products are Transforming Companies. *Harvard Business Review*, 1–19.

Project Management Institute, Inc. (2008). *A Guide to the Project Management body of Knowledge (PMBOK Guide)-Forth Edition*. Newtown Square, PA: PMI Publications.

Reim, W., Parida, V., & Örtqvist, D. (2015). Product-Service Systems (PSS) business models and tactics - a systematic literature review. *Journal of Cleaner Production*, *97*, 61–75. doi:10.1016/j.jclepro.2014.07.003

Ross, J., Sebastian, I., & Beath, C. (2016). How to Develop a Great Digital Strategy. *MIT Sloan Management Review*.

Shenhar, A., & Dvir, D. (2007). *Reinventing project management: The diamond approach to successful growth and innovation*. Boston: Harvard Business Press.

Shenhar, A., Holzman, V., Melaned, B., & Zhao, Y. (2016). The Challenge of Innovation in Highly Complex Projects: What Can We Learn from Boeing's Dreamliner Experience? *Project Management Journal*, *47*(2), 62–78. doi:10.1002/pmj.21579

Shirky, C. (2008). *Here Comes Everybody: How Change Happens when People Come together*. London: Penguin Books.

Slama, D., Puhlmann, F., Morrish, J., & Bhatnagar, R. (2015). *Enterprise IoT*. O'Reilly Media, Inc.

Stacey, R. (1995). The science of compelxity: An alternative perspective for strategic change processes. *Strategic Management Journal*, *16*(6), 477–495. doi:10.1002mj.4250160606

Sting, F., Loch, C., & Stempfhuber, D. (2015). Accelerating Projects by Encouraging Help. *MIT Sloan Management Review*, 5–13.

Svejvig, P., & Andersen, P. (2015). Rethinking project management: A structured literature review with a critical look at the brave new world. *International Journal of Project Management*, *33*(2), 278–290. doi:10.1016/j.ijproman.2014.06.004

Tanoto, Y., & Setiabudi, D. (2016). Development of autonomous demand response system for electric load management. In *2016 Asian Conference on Energy, Power and Transportation Electrification, ACEPT 2016* (pp. 1-6). Marina Bay Sands: IEEE. 10.1109/ACEPT.2016.7811532

Teece, D. (2007). Explicating dynamic capabilities: The nature and microfoundations of (sustainable) enterprise performance. *Strategic Management Journal*, *28*(13), 1319–1350. doi:10.1002mj.640

Van Alstyne, M., Parker, G., & Choudary, P. (2016, April). Pipelines, Platforms, and the New Rules of Strategy. *Harvard Business Review*, 54-60, 62.

Wang, W., He, Z., Huang, D., & Zhang, X. (2014). Research on Service Platform of Internet of Things for Smart City. In J. Jiang, & H. Zhang (Ed.), *ISPRS Technical Commission IV Symposium. XL-4* (pp. 301-303). Suzhou: Int. Arch. Photogramm. Remote Sens. Spatial Inf. Sci. 10.5194/isprsarchives-XL-4-301-2014

Weill, P., & Woerner, S. (2015). Thriving in an increasingly digital ecosystem. *MIT Sloan Management Review*, *56*(4), 27–34.

Weiser, M. (1991). The computer for the twenty-first century. *Scientific American*, *290*, 46–55.

Willness, C., & Bruni-Bossio, V. (2017). The curriculum innovation canvas: A design thinking framework for the engaged educational entrepreneur. *Journal of Higher Education Outreach & Engagement*, *21*(1), 134–164.

Winter, M., Smith, C., Morris, P., & Cicmil, S. (2006). Directions for future research in project management: The main findings of UK government-funded research network. *International Journal of Project Management*, *24*(8), 638–649. doi:10.1016/j.ijproman.2006.08.009

Woo, C., Jung, J., Euitack, J., Lee, J., Kwon, J., & Kim, D. (2016). Internet of Things Platform and Services for Connected Cars. In M. Ramachandran, G. Wills, R. Walters, V. Mendez Muñoz, & V. Chang (Ed.), *Proceedings of the International Conference on Internet of Things and Big Data* (pp. 469-478). Rome: Science and Technology Publications, Lda. 10.5220/0005952904690478

KEY TERMS AND DEFINITIONS

Collaborative Culture: It is teamwork based on trust, communication, and a shared vision or purpose.

Competitive Advantage: Competitive advantage results from a firm's ability to perform the required activities at a collectively lower cost than rivals, or perform some activities in unique ways that create buyer value.

Data Analytics: A process that uses quantitative and qualitative techniques to uncover knowledge from raw data with the aim of generating value.

Diamond of Innovation: A framework for project classification based on novelty of the product, extent of new technology to the organization used by the project, complexity of the product, and project urgency.

Digital Ecosystem: An ecosystem is composed of consumers, producers, and competitors, and as a result of their interactions over a platform driven by the ecosystem leader, innovations (planned or not anticipated) emerge.

Digital Transformation: A digitization resulting from the combination of mobile, cloud computing, social, big data, and internet of things technologies with impact on organizations and society.

Management of a Business Network: The management of network including several firms, where the firms engage from time to time in mutual projects.

Management of a Project Network: The management of a network including several firms that are participating in a project.

Portfolio Project Management: The belief that family is central to wellbeing and that family members and family issues taking precedence over other aspects of life.

Product-as-a-Service Business Model: Product-as-a-service blends physical products, accompanying services, and monitoring software where the buyer may no longer own a physical thing.

Chapter 4

Smart Cities, Smart Grids, and Smart Grid Analytics:
How to Solve an Urban Problem

Shaun Joseph Smyth
Ulster University, UK

Kevin Curran
Ulster University, UK

Nigel McKelvey
Letterkenny Institute of Technology, Ireland

ABSTRACT

The introduction of the 21st century has experienced a growing trend in the number of people who choose to live within a city. Rapid urbanisation however, comes a variety of issues which are technical, social, physical and organisational in nature because of the complex gathering of large population numbers in such a spatially limited area. This rapid growth in population presents new challenges for the already stretched city services and infrastructure as they are faced with the problems of finding smarter methods to deal with issues including: traffic congestion, waste management and increased energy usage. This chapter examines the phenomenon of smart cities, their many definitions, their ability to alleviate the discomforts cities suffer due to rapid urbanisation and ultimately offer an improved and more sustainable lives for the city's citizens. This chapter also highlights the benefits of smart grids, their bi-directional real-time communication ability, and their other qualities.

DOI: 10.4018/978-1-5225-3996-4.ch004

Copyright © 2018, IGI Global. Copying or distributing in print or electronic forms without written permission of IGI Global is prohibited.

INTRODUCTION

Urbanism has existed for more than 5,000 years witnessing cities being formed according to their landscape, their position from the sea, the ruling of rivers and the transportation networks that connect cities (Anthopoulos & Vakali, 2012). The period between 1950 and 2005 has witnessed a marked annual increase in urbanisation within the developing countries, with a reported growth rate of 3.6% compared to the 1.4% experienced in industrialised countries. By the year 2000 an estimated 45% of the population in developing countries (1.97 billion) and 75% (945 million) of developed countries were already living in cities (Khansari et al., 2014).

Between the years 2009 and 2050 it is predicted that the world's population will increase by 2.3 billion taking it to 9.1 billion people in total and over this same period the population within urban areas is predicted to grow by 2.9 billion to reach 6.3 billion people or 70% of the world's population by 2050. These figures hide a significant difference between the new and emerging markets with the least developed countries witnessing the most dramatic population growth and urbanisation (Bélissent, 2010).

The United Nations (UN) has predicted that by the year 2050 the rapid relocation to cities will have caused the world's urban population to increase by 75% and the result of this migration to the cities will be an increase in the number of densely populated areas (Barrionuevo et al., 2012). This figure differs as a 2007 to 2008 United Nations World Urbanisation Prospects study claims that the population within urban areas is to gain 3.1 billion surpassing the 3.3 billion in 2007 to a figure of 6.4 billion in the year 2050 as shown in Figure 1 (Washburn et al., 2009).

Cities were formed as a natural response to changed life circumstances and have also had a profound and lasting impact on the further development and progress of the human species (Schuurman et al., 2012). Cities are the future of humankind. The 18th century witnessed less than 5% of the World's population living in a city and a huge majority of these were simply engaged in generating enough food to live. The entry into the 21st century however, has been accompanied with a strong worldwide inclination to increase the concentration of the population within fairly few large cities. These large dense cities are attractive and appeal to their citizens as they have the potential to be both highly productive and pioneering and thus very attractive for our futures (Harrison & Donnelly, 2011). Despite only making up 2% of the world's surface cities house half of the world's population, consume 75% of our energy resources, and produce 80% of the carbon which is harming our environment (Aoun, 2013).

Figure 1. Urbanisation causing cities to outgrow rural areas (Washburn et al., 2009)

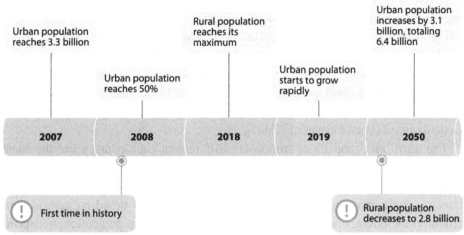

Urbanisation is driven by profound technological, social and economic changes which are incurred because of globalisation processes which in turn are forcing cities globally to adopt ground-breaking, competitive, sustainable and long-term policy strategies (Schuurman et al., 2012). Urban living is beginning to adopt a principal role in how humankind evolves as today there are reportedly more than 1 in 2 living in urban environments (Dohler et al., 2011).

As people worldwide continued to move into cities in greater numbers in a bid to improve their economic circumstances city populations continued to increase to the point where the period between 1990 and 2000 witnessed urbanisation in developing regions marked by the appearance of cities which prior to 1990 did not exist. However, along with this rapid growth and the increase in numbers migrating to the cities, the increased urbanisation is posing many major challenges to our societies including air pollution, overconsumption of resources, city congestion along with urban insecurity (Steinert et al.,2011; Hall, 1988). City governments are faced with these many challenges and others such as higher crime rates, difficulties in waste management, wasteful and increased energy consumption, the basic delivery of public services, and so on (Alawadhi et al., 2012; Steinert et al., 2011).

Rapid urbanisation increases the challenges within cities and the United Nations (UN) estimates that between 2008 and 2009 the population in both rural areas and urban areas became equal, while by the start of 2019 the number of people living in cities will exceed the numbers residing in rural areas. As the numbers residing in cities increases energy shortage presents itself as a major problem and the U.S. Energy Information Administration (EIA), an agency within the U.S. Department

105

of Energy anticipates that the World marketed energy consumption will increase by 44% from 2006 to the 2030's. The EIA also anticipates that energy demand from emerging economies such as the BRIC countries (Brazil, Russia, India and China) will grow by 73% in this same time frame which far exceeds the 15% increase from developed countries such as the United Kingdom, the United States, Australia, France, Japan and Germany. Increases in price are common occurrences and caused by the high costs of developing new generation capabilities, the directives for more secure and renewable energy forms and the tendency for countries to import energy, particularly oil (Washburn et al., 2009).

The term 'city' and its definition are different in each country but the most common definition refers to a relatively large and permanent settlement. Cities have highly dense populations and their inhabitants mainly live because of work within industry, commerce and services. Cities are operationally based on several core structures including: energy, water, information and telecommunication, transport, business market, city services, citizens and sanitation (Morvaj et al., 2012).

In today's world where current resources are already threatened it is important to make cities greener and ultimately more sustainable as urban areas are consuming many of the resources which are currently available. Advanced systems that will enable processes to be both automated and improved inside a city will have a key function within smart cities. From the smart design of buildings which retain the rain water for later use to the intelligent control systems that monitor infrastructures autonomously, the potential improvements which are enabled by sensing technologies are immense. With predictions highlighting that the global economy will be considerably unbalanced due to the expected growth of cities, it is forecasted that by 2050 urban areas will exceed 6 billion people and such growth will further exacerbate the existing climate and energy challenges that urban areas already experience. In a bid to address these challenges cities which are more resource efficient and technology driven are required and the smart city promises to take advantage on its economic opportunities and social benefits as it alleviates the pains of urbanisation (Hancke & Hancke Jr, 2012; Washburn et al., 2009).

Several other concepts such as the smart city have all been developed and put in place such as the digital cities and ubiquitous cities in a bid to achieve both a competitive and a sustainable influence, as the cities themselves firmly believed that innovative uses of ICT would nurture sustainable city innovation which can improve the quality of life of its citizens (Schuurman et al., 2012).

One approach adopted by several cities in a bid to curb the urban planning challenges was to transform the city into a smart city (Kuyper, 2016). As a result, momentum within the European Union (EU) continued to grow as almost every European city has witnessed various projects being started as attention to smart cities increases (Schuurman et al., 2012).

With the constant rise in world population causing an increased consumption of resources and leading to resource shortages and climate change the need for ground-breaking solutions is clear to see. Urban areas are mainly responsible for the shortage in resources thus prompting a growing need to create smart infrastructures in a search of more greener and energy efficient urban dynamics. Solutions to these problems involve improvements to most of the components of urban dynamics and ultimately the smart city itself as displayed in Figure 2.

The smart city concept can be viewed as a recognition of the increasing importance digital technologies for both achieving a competitive position and sustainable future however. The smart city-agenda grants ICTs the task of providing a vision of how to alleviate and resolve the challenges associated with rapid urbanisation through achieving many of the following goals: reduction in energy consumption, reduced carbon emissions, sustainable growth, improved quality of life for its citizens and ultimately a better urban life (Steinert et al., 2011; Schuurman et al., 2012).

For several years now cities have been incorporating the use of new technologies however, lately the rate at which technology is currently being implemented is on the increase. Cities have been making use of new technologies for several years now and as the degree of technology adoption increases cities around the world are ultimately becoming smarter. Newer technologies along with faster and easier connectivity enables cities to optimize resources, save money, and at the same time provide better services to its citizens (Cerrudo, 2015).

Figure 2. Sensing in Smart Cities (Hancke & Hancke Jr, 2012)

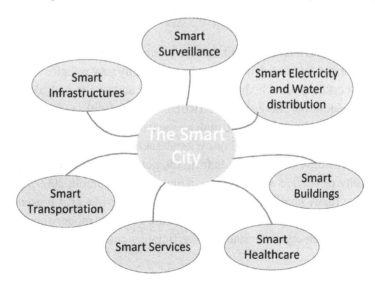

With recent years, having witnessed the number of world inhabitants residing in cities surpassing those residing in the countryside and the population of the Earth measuring 7170 million as of July 2014, there is a huge demand on the quantity of energy required for this number of individuals to carry out the tasks in their daily lives (Rodríguez-Molina et al., 2014).

There has however been no change to the basic structure of the electrical power grid for the last 100 years even though the average age of power grade transmission lines in the U.S. are exceeding 50-60 years. The U.S. Department of Energy (DOE) has stated that the past 20 years has experienced an increase in both demand and consumption of 2.5% annually. Experiences have shown that the hierarchical centrally controlled grid of the 20th Century is ill equipped to cater for the needs of the 21st Century as additional challenges and issues such as power-grid integration, system stability and energy storage also require addressing. Deficiencies and issues in the present system such as congestion, slow response times and safety related factors have all contributed to blackouts over the last 40 years (Gungor et al., 2010; Gungor et al., 2011). The fact that there has been a total of three major blackouts within the last ten years and it is the dependence upon on old technology which leads to uneconomical systems which are costing needless amounts of money to the utilities, consumers, and taxpayers (Metke & Ekl, 2010).

January 2009 witnessed the U.S. Department of Energy (U.S.DOE) release an assessment on the state of the U.S. electricity grid which was based on the work carried out by its Electricity Advisory Committee (EAC). The report which was officially titled "Keeping the Lights on in a New World". This report paints a very bleak picture of a Grid which is neither modern nor smart highlighting that much of the electricity supply and delivery infrastructure is approaching the end of its useful life (Collier, 2010).

With recent years has witnessed that the ageing U.S. power grids have become both underinvested, overburdened and subject to many new policies and challenges including uncertainty in schedules and transfers across regions along with the infiltration of renewable energy systems (RES). It is faced with the unpredictability of events due to limited knowledge and management of complex systems and the threat of terrorist attacks, either physical or cyber-attacks. The number of consumers has also increased as have their demands from the power system as they are expecting a better quality of service and a more reliable supply (Momoh, 2009).

Due to the many challenges, which, the existing power grid presents the new concept of the smart grid which is based on a more solid and modern communication infrastructure has emerged for enhanced efficiency and reliability (Gungor et al., 2011).

Electricity demands worldwide continue to rise with the ever-increasing population of the Earth and the smart grid presents itself as a very attractive and convincing system for the future of energy especially as the Earth has abundant, yet limited resources and it is important to justify the use of energy and allow the usage of renewable energies, thus providing electricity to use with a lower impact on the planet. The smart grid meets these needs as it not only combines efficient energy consumption with new and innovative technologies related to renewable energies, but it can also provide numerous valuable services including data provision and power monitoring (Rodríguez-Molina et al., 2014).

Both Smart Grids and Smart Cities involve the use of advanced electrical engineering and service technologies which are both assisted by ICT and additional solutions to efficiently manage complex infrastructure systems. Significant interactions exist between both Smart Grids and Smart Cities with the shared principles including intelligent interconnectivity, integration including the end user element of the Smart Grid which is also an important element of the Smart Cities (Forfás, 2013).

BACKGROUND

Smarter cities refer to those urban areas that utilise operational data such as that which arises from city traffic congestion, power consumption and public safety events to enhance the operational properties of city services (Harrison et al., 2010). With the world's urban population continuing to increase many governments are contemplating adopting the smart city concept within their cities. This involves the implementation of big data applications which will support smart city components enabling them to reach the required level of sustainability and thus improving living standards. Smart cities exploit multiple technologies to improve the performance of health, transportation, energy consumption, education, and water services leading to higher levels of comfort of their citizens. Achieving this involves cities reducing both costs and resource consumption as well as engaging with their citizens more effectively (Al Nuaimi et al., 2015).

Two of the many cities which have embraced this concept, tackling the challenges presented by urban planning and converting to smart cities are Barcelona and Amsterdam with both continuing to invest in smart city strategies to this day (Kuyper, 2016).

A recent technology which has an enormous potential in the enhancement of smart city services is big data analytics. Digitization has become an essential part of everyday life and the collection of data has resulted in the amassing of large amounts of data which can be used in many beneficial application areas. A crucial factor for success within many businesses and service domains which includes the smart city area is the effective analysis and utilisation of big data (Al Nuaimi et al., 2015).

The concept of the smart city encompasses and understands the growing importance of Information and Communication Technologies (ICT). It is vital that strategies are quickly identified and related actions performed to make cities smarter, i.e., more operationally effective, sustainable from an environmental perspective and performing in a cost-effective manner. To meet these goals smart cities, need to be managed, measured and monitored in an intelligent manner. The term "smart" is frequently replaced by other adjectives such as interconnected or intelligent. It is through the automation of services, buildings, traffic systems etc and the technological advancements in data collection that we can monitor, understand analyse and subsequently plan cities in such a way as to improve their efficiency and ultimately the quality of life for its citizens (Carli et al., 2013).

THE SMART GRID

The term grid was traditionally used for an electricity system which would support all four or some of the following operations: electricity generation, electricity transmission, electricity distribution and finally electricity control (Fang et al., 2012). Smart grids comprise both hardware and software tools which allows electricity generators to transmit power more efficiently, decreasing peak capacity requirements and permitting real-time, interactive exchange of information with customers. Globally, smart grid technologies also display the potential to reduce carbon emissions (Steinert et al., 2011).

Electric systems worldwide however, have encountered many challenges including an ageing infrastructure, reliability of the electrical system, security issues which threaten the development of power systems, a continuing increase in energy demand, an increase in the number of renewable energy sources (RES) and the need to provide a secure supply of electricity to the consumer as the conventional electricity distribution systems that exist are primarily non-intelligent as these energy distribution structures only deliver a unidirectional flow from the generating station to the consumer with the supply of electricity either estimated or predicted using previous available data One improvement to the existing distribution scheme is to employ a bi-directional system which allows the flow of electricity to a client's premises and vice versa. This enables a more resourceful use of energy allowing electricity to flow back to the utility to be stored for later use in instances of low demand and such systems are collectively known as smart grids. While the traditional unidirectional power grid system has served well for the past century the modern society of today now requires a more reliable system which is more scalable, economical, manageable, cost effective and environmentally friendly. It is due to these challenges that the next generation electric power system was born. The concept of the Smart Grid has gained increased attention since its proposal by the U.S. Electric Power Research Institute in 2001 and has been propelled even further by the promotion of low carbon economies in developing countries (Yu et al., 2012; Hancke & Hancke Jr, 2012; Bari et al., 2014).

One of the critical features of the Smart Grid is use of both information and communication technologies to collect and act on information in an automated manner to increase the reliability, efficiency, sustainability and the cost effectiveness of the production, transmission and distribution of electricity. The standard Smart Grid comprises several power-generating entities and power consuming entities which are all connected through the means of a network Grid. Figure 3 displays an example of the communication architecture which exists within the Smart Grid as the generators feed energy into the grid and the consumers draw energy from the grid (Bari et al., 2014).

Using a two-way digital technology, the Smart Grid can deliver power to the consumer or end user enabling a more efficient management of the consumer's end uses of electricity and a more efficient use of the grid to correctly identify and adjust supply demand-imbalances promptly and detect faults through the self-healing process thus improving the quality of service provided, enhancing reliability of the grid and reducing costs. The Smart Grid involves more than simply installing smart

Figure 3. Example of communication architecture in the Smart Grid (Bari et al., 2014)

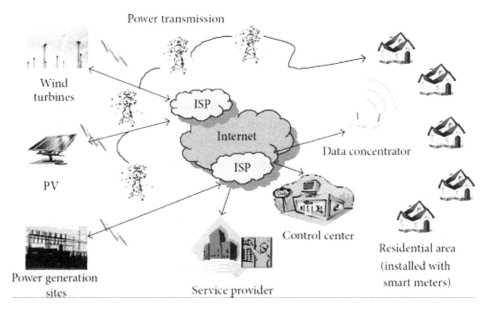

meters as the introduction of information technology to the electric grid develops various applications that use the devices networking and communications technology capabilities and control and data management systems (Bari et al., 2014).

One such application in Smart Grid communication and a key task in the Smart Grid is the Advanced Metering Infrastructure (AMI). Unlike the old-style method which involved technicians physically visited each consumer monthly to record the data manually the Smart Grid's two-way communication ability enables smart meters in AMI to provide real-time monitoring of power activities and consumer usage to be collected on a periodical basis (e.g. every 15 minutes) by a data concentrator using either wired or wireless communications and forwarded to a central location. Such real-time data is more efficient and precise and allows reports on power consumption to be created which enables (a) the achievement of balance of power demand and supply, (b) analysis of consumer energy usage, (c) historical data on energy consumption, (d) billing information, (e) pricing and (f) suggestions on reducing peak load (Bari et al., 2014; Li et al., 2012).

Electricity is currently the fastest-growing component of the total global energy demand. The requirement to meet this high demand for energy worldwide in a cost effective, secure and sustainable manner is motivation for investment in this area the further development of a high growth market for smart grids. One estimate places the total global investment in smart Grids between 2008 and 2015 at US$200 billion while a study for the EU Commission predicts that the total investment in the EU, U.S. and China will be a total of 365 billion euros by 2020. The market for Smart Grid technology infrastructure which includes smart meters, sensor networks, fibre optic and wireless networks, data analytics is also very lucrative and expected to rise to almost US$ 16 billion by 2020 (Forfás, 2013).

Within the electric utility industry, the Smart Grid is viewed as an up and coming business strategy as there is a need to reduce energy consumption through energy efficiency and demand response. To meet rising demands there is a need for utilities to increase generation capacity with generation coming from a mix of oil, gas, nuclear and green power and consumers are also asked to bear the burden by reducing their demand using new and efficient appliances or installing smart home controls (Roncero, 2008). As the demand for electricity as a global energy continues to grow there is in comparison to the conventional power grid, a growing demand for a sustainable, secure, clean and cost effective solution which offers a two-way flow of electricity and information between the supplier and the consumers of electric power and promotes the use of renewable energy sources (RES). Subsequently there is a growing market for the Smart Grid which involves a series of new technologies and offers these qualities in the future power grid development (Gharavi & Ghafurian, 2011; Saxena & Choi, 2015; Hancke & Hancke Jr, 2012; Yu et al., 2012; Hernández et al., 2012).

The Smart electricity Grid opens the door to many new and different applications with widespread impacts: providing the capability to incorporate more renewable energy sources (RES), electric vehicles and distributed generators into the network, ensuring the reliable and efficient delivery of power through demand response and thorough control and monitoring capabilities, the use of automatic grid reconfiguration to prevent or restore power outages (self-healing capabilities) allowing consumers to have better control over their electricity consumption and to enable them to actively participate in the electricity market (Giordano et al., 2011). A secure and reliable delivery of power around the country is achieved using a clever monitoring system which not only keeps track of energy coming in from various sources but the two-way bidirectional communication system collects data about how and when consumers are using power and can detect where energy is required surpassing the existing one directional system of power supply (Ling et al., 2012).

DEFINING THE SMART GRID

The term grid was traditionally used for an electricity system which would support all four or some of the following operations: electricity generation, electricity transmission, electricity distribution and finally electricity control (Fang et al., 2012).

Due to the complex nature, of power systems it is difficult to place a definitive definition or description on the Smart Grid as this differs depending upon location as does the vision of the stakeholders and the technological complexities involved. The European Regulators' Group for Electricity and Gas (ERGEG) defines the Smart Grid as an electricity network which efficiently incorporates the performance and activities of all the users which are connected to it such as the generators or the consumers and those that do both in a bid to ensure an economical and efficient power system which has low losses and high levels of quality and security of supply and safety (Yu et al., 2012; Bari et al.,2014).

The U.S. department of Energy (DOE) as highlighted by (Yu et al., 2012; Ardito et al., 2013) assigns a more detailed definition to the Smart Grid highlighting that the functions that the Grid must provide include the following:

- A self-healing quality from power disturbance events.
- Allow the active participation by consumers in demand response.
- Operate robustly against both physical and cyber-attacks.
- Provide high quality power suitable to meet the needs of the 21st century.
- Support for different types of storage and power generation.
- Enabling new products, services and markets.
- Enhancing assets and operating in an efficient manner.

The U.S. department of Energy (DOE) has also put forward the following definition for the Smart Grid:

An automated, widely distributed energy delivery network, the Smart Grid will be characterized by a two-way flow of electricity and information and will be capable of monitoring everything from power plants to customer preferences to individual appliances. It incorporates into the grid the benefits of distributed computing and communications to deliver real-time information and enable the near-instantaneous balance of supply and demand at the device level (Bari et al., 2014).

The definition of the Smart Grid in China differs again with the perception concentrating on all aspects of the power system to include smart power generation, transmission, distribution, storage and consumer usage. The Smart Grids in China are therefore subsequently defined as an integration of renewable energy, intelligent decision making, the adoption of new materials, energy storage technology, advanced equipment, information technology, control technology all of which can recognise digital management, intelligent decision making and interactive transactions in the generation of electricity, transmission, distribution, usage and storage (Yu et al.,2012).

KEY FEATURES AND REQUIREMENTS OF THE SMART GRID

The Smart Grid has several key requirements and characteristics and a list of the most relevant achievements and major functions include the following:

1. **Self-healing:** The Grid has a reduction in restoration time and maintenance due to predictive analytics and the self-healing attribute of the Grid which is aware of the status of every major component in real or near real time and has control equipment to provide optional routing paths which provides the capability for alternative flow of electricity throughout the system to maintain power to all customers (Gharavi & Ghafurian, 2011; Ardito et al., 2013; Roncero, 2008; Giordano et al., 2011).
2. **Secure Electricity Supply:** Due to the two- way end to end communication capability which the Smart Grid supplies there is the need for secure communications as the requirement of both physical and cyber security of all assets is critical (Gharavi & Ghafurian, 2011; Ardito et al., 2013; Saxena & Choi, 2015; Yu et al., 2012; Kim et al., 2015
3. **Interactive:** The Smart Grid integrates both electricity and communication in an electric network which supports a new generation of interactive energy, supplying the end consumer with digital quality electricity after making

intelligent decisions regarding the transactions of electricity, generation, transmission, deployment usage and storage (Gharavi & Ghafurian, 2011; Roncero, 2008; Yu et al., 2012).

4. **Predictive:** The Smart Grid uses machine learning, impact projections and analysis to make predictions of the next most likely events so that appropriate actions can be taken to reconfigure the system. Through the Smart Grid's exchange of information between different elements it can deliver predictive information and advice to utilities, their suppliers and their customers also on how to manage their power supply in a better manner. The self-healing and predictive qualities of the Grid also reduces restoration time (Roncero, 2008; Gharavi & Ghafurian, 2011; Kim et al., 2015).

5. **Renewable Energy:** The Smart Grid through two-way communication can increase the ability to deploy the addition of renewable energy sources (RES) to assist with addressing global climate change and the integration of distributed and renewable energy provides electric power to consumers in a more reliable and efficient manner (Saxena & Choi, 2015; Gharavi & Ghafurian, 2011; Yu et al., 2012; Giordano et al., 2011).

6. **Sustainable:** The Smart Grid enables better use of the assets, it should provide long term sustainable power system which is aware of all the actions of the users connected to it to enable it to efficiently provide a sustainable supply of electricity with low losses and high levels of quality and security for the safe supply of power (Ardito et al., 2013; Gharavi & Ghafurian, 2011; Yu et al., 2012).

7. **Economic:** In comparison to the traditional power grids the Smart Grids are both economically efficient and environmentally friendly (Yu et al., 2012).

8. **Cleaner Energy:** The introduction of the Smart Grid should allow for the use of electric vehicles thus reducing the need for hydrocarbon fuels resulting on a reduction of the carbon footprint of modern society and an increase in low carbon economies within developing countries (Yu et al., 2012; Gharavi & Ghafurian, 2011).

Development of the Smart Grid and the key features and requirements as categorised above does not involve replacement of the current existing electricity network as such a procedure would be impossible both from a technical and economical perspective. Development of the Smart Grid is however an enhancement of the existing network through the introduction of new features and services. The Smart Grid however continues to build on the available infrastructure increasing the employment of existing properties and empowers the application of the new functionality (Ardito et al., 2013; Gharavi & Ghafurian, 2011).

POLLUTION

The current rates at which the resources of the Earth are being consumed are unsustainable and as such is creating a major environmental problem. Climate change, depletion of resources and air pollution have a major influence on both the Earth and its population resulting in a need to change our current behaviour (Boudreau et al., 2008). The rapid urbanisation which has been experienced has had an environmental impact even though cities only account for a mere 2% of the planet they are however, accountable for 60 to 80 percent of energy consumption and 75% of carbon dioxide emissions which contribute to global warming. (Boudreau et al., 2008; Barrionuevo et al., 2012; Albino et al., 2015; Aoun, 2013; Kennedy et al., 2009).

While current technology such as coal-fired power stations provides, the essential electricity required to support our comfortable lifestyles they also create carbon emissions and contribute to global warming at the same time (Boudreau et al., 2008). As electricity consumption leads to 40% of worldwide carbon dioxide emissions mainly since almost 70% of electricity is currently produced from fossil fuels there is a need for smart grid technologies to reduce these emissions through feedback on current energy usage, more efficient usage and improved peak load management (see (https://www.iea.org/publications/freepublications/publication/smartgrids_roadmap.pdf).

As public concern over environmental issues persists in growing, worldwide efforts for a greater environmental sustainability have not been able to stabilise or even decrease the atmospheric level of greenhouse gases (GHGs). The largest percentage (49%) of GHG emissions are produced by the energy supply sector and due to an increased energy demand and a larger share of coal in the global fuel mix, GHG emissions associated with the energy supply sector have increased more rapidly between 2000 and 2010 than in the previous three decades with the year 2013 witnessing the levels of carbon dioxide, the principal GHG produced due to the burning of fossil fuels such as coal and oil surged to its fastest rate in 30 years with a total of 50 billion tons of carbon dioxide being emitted annually (Kranz et al., 2015).

With the increasing demand for electricity there has also been a rapid rise in electricity consumption- related carbon dioxide emissions. In all areas concerning carbon dioxide emissions a total of 40% can be attributed to the electricity industry in China. As the demand for electricity has increased the electricity consumption related carbon dioxide emissions have also increased (Figure 4). With both economic output and energy consumption continuing to rise in the future there is a major need to develop a power system which has both high efficiency and low carbon emissions for the healthy development of the electricity industry (Yu et al.,2015).

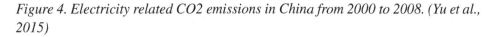

Figure 4. Electricity related CO2 emissions in China from 2000 to 2008. (Yu et al., 2015)

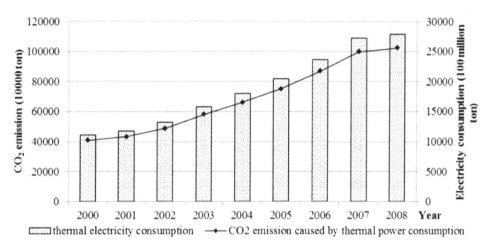

One of the main goals of smart cities is ensuring that the level of pollution remains at an acceptable level by paying attention to carbon emissions which present a serious threat to both our planet and health. With rapid urbanisation comes increased traffic pollution, waste and energy costs all of which present as threats to human health and sustainability (Barrionuevo et al., 2012; Hancke & Hancke Jr, 2012).

TRANSPORTATION

Smart transport is focussed on the development of infrastructure, technologies and systems all of which are energy efficient, have low impact on the environment and provide the necessary mobility required by society in a cost-effective manner. Research into the area of intelligent transport systems is receiving much interest and investment from leaders in telecommunications and software and leading car manufacturers (Dohler et al., 2011). Urban planners can also use of big data to improve decisions made regarding both road and mass transit traffic patterns to better plan for future developments or improvements to the infrastructure (Hiller & Blanke, 2016).

Facilitating methods by which citizens can get around cities and access public services is a major challenge for urban planners especially since the increase in urban populations. However, many cities are already addressing this issue such as Curitiba, Brazil which has introduced a transportation system to improve mobility for which a fleet of 2,160 buses have already been designated providing the city with high capacity, high speed, high frequency buses complemented by other lines operating between neighbourhoods. The inclusion of an extra120 kilometres of bike lanes within the city also displays the city's movement towards tackling mobility for the city's citizens (Barrionuevo et al., 2012).

Work on the other side of the world in Singapore is being carried out on a pilot project to ascertain the most effective and beneficial technology for enabling urban mobility and the city has already installed a system of traffic sensors enabling authorities to predict traffic jams up to one hour in advance (Barrionuevo et al., 2012).

The transport sector however, is also the second leading source of worldwide GHG emissions after energy. ICT solutions have the ability to reduce transport needs by making it easier to mix methods of transportation and select the most energy-efficient type of transport. Such solutions can also help optimise routes and reduce inventory needs encouraging more efficient driving. With the increase in fuel prices many companies will adopt the use of the more energy efficient ICT solutions which in turn has a major influence on the reduction of emissions (Steinert et al., 2011).

Of all the energy, related activities which humans carry out in an urban environment traffic is one of the most energy expensive ones. In addition, it is also characterised by large waste as only 30% of the fuel potential is used to transfer kinetic energy to the actual vehicle with the majority of this energy is dissipated in decelerating phases by brakes and gases. Power bumps are however, pioneering energy harvesting devices which reduce the speed of vehicles through the conversion of kinetic energy which is otherwise wasted by brakes into electricity. For evident reasons of energy balance these devices should not be randomly placed on road networks but however positioned in decelerating sites only which includes urban road crossings, pedestrian crossings, main road exits and where standard passive speed bumps are installed (Pirisi et al., 2012).

The large increase in car ownership in recent years highlights the need for better traffic management top avoid the occurrence of traffic jams and improve the flow of traffic especially at intersections. A traditional way to regulate and monitor the flow of traffic involves using traffic lights. Typically, these have fixed switch interval times (*i.e.*, from red to green and to yellow), and this is not altered according to traffic conditions. Traffic jams have serious influences on fuel consumption due to the increased frequency in starts and stops and there is also increased carbon emissions as a result. Estimation of the number of cars approaching an intersection would generate useful information whereby switch interval times could be adjusted based on the volume of traffic. For such a system to be realised the detection of traffic and the counting of the number of cars is required. The removal of stop signs and traffic lights to tackle traffic conditions has also been proposed however, this requires vehicles which are connected wirelessly and able to communicate with each other to adopt the use of decision making for collision avoidance and using a time slot technique a vehicle would be able to negotiate intersections with wireless communication taking place through RFID (Hancke & Hancke Jr, 2012).

The ownership of electric vehicles is a growing concern and continues to be more popular as environmental concerns increase (Moslehi & Kumar, 2010). Electric vehicles are a means of reducing the reliance on fossil fuels which in today's world are the principal energy source in both the transport and power generation industry. Depletion of these resources which are unsustainable has led to the need for an alternative solution as the oil economy is both unsustainable and very limited. The burning of fossil fuels produces emissions of greenhouse gases (GHGs) which are highly influential in world climate change (Moslehi & Kumar, 2010; Mwasilu et al., 2014).

By 2050 due to the significant increase in electric vehicles (EV) and plug-in hybrid electric vehicles (PHEV) the transport sector will account for 10% of overall electricity consumption. If vehicle charging is not managed correctly it could increase the peak demands experienced in both residential and service sectors thus requiring infrastructure investment to avoid supply failure. Over the long term the smart grid technology could allow electric vehicles to feed electricity stored in their batteries back into the system when required. (see (https://www.iea.org/publications/freepublications/publication/smartgrids_roadmap.pdf).

RENEWABLE ENERGY SOURCES (RES)

Even though fossil fuels are the main source of energy worldwide today, the world is however considering the use of alternatives to energy generation systems which are more economical and environmentally friendly. There are several factors which influence this decision and they include: (a) an increased demand for electric power from both developing and developed countries, (b) the lack of resources within many developing countries to build power plants and distribution networks, (c) some industrialised countries facing insufficient power generation and (d) greenhouse gas emissions and concerns over climate change (Mohd et al., 2008).

The U.S. accounted for 95% of the installed wind energy worldwide in the early 1980's but recorded figures saw this figure drop to only 22% in 1998 as other nations adopted this practice of power generation. The past 30 years has witnessed the rapid growth in sales of both solar and wind power systems as the cost of the electricity generated have continued to improve while investment costs decline (Herzog et al., 2001).

The market for energy storage technologies has also witnessed a rapid growth and Ireland has significant advantages for research, development and the trialling of such technologies as Ireland's island status presents major early challenges for the grid system having limited interconnections with its neighbours and boasts a rich wind energy source with a target of 40% of electricity to come from renewable sources by the year 2020 which is the highest in the EU for variable renewables within a single system (Forfás, 2013).

The Smart Grid through its ability to allow a two-way flow of electricity and information is however expected to meet the required needs of the modern power system and revolutionise electricity generation, transmission and the distribution process. The Smart Grid also complements the current electric grid by the inclusion of renewable energy sources (RES) such as wind, solar and biomass which are more environmentally friendly and cleaner in comparison to the fossil fuels currently used in many bulk electric power generation facilities in the majority (Bari et al., 2014). While centralised sources of generation will continue to play a major role within the Smart Grid, large-scale wind and solar generation both however, also have a part to play in the generation of power while the cost is justified (Gharavi & Ghafurian, 2011).

The impact of human factors on climate change has forced the scientific community to focus on renewable energy systems as they represent a way of maintaining our current way of life and energy consumption levels without depleting our non-renewable sources such as carbon and fossils which are currently the primary cause of climate change (Pirisi et al., 2012).

Renewable energy sources have an enormous potential as they can in principle meet the world's energy demands many times over and the provision of sustainable energy services can be achieved using normal, accessible, renewable energy sources such as biomass power plants, fuel cells, wind turbines, solar-thermo power, hydropower turbines, photovoltaic solar systems, gas micro-turbines, combined heat and power (CHP), micro turbines, and hybrid power systems. Wind has enormous potential as it is a globally clean source of energy which is available worldwide and produces no pollution during power generation (Herzog et al., 2001; Mohd et al., 2008). With recent advances in the technology of power systems, the integration of wind energy into the Smart Grid has become a reality and has seen both on land and offshore wind energy becoming major workings of the energy systems worldwide. The Smart Grid enables the connectivity of wind turbines as occasional sources of energy while advanced turbines with power electronics and other devices are authorised to support the Grid with reactive power protecting equipment during severe Grid disturbances (Gharavi & Ghafurian, 2011).

Renewable energy sources including wind, biomass, solar geothermal and hydropower can all provide renewable energy sources based on the use of regular local resources which are all routinely available. Their transition into renewable based energy systems is becoming more attractive as their costs have dropped as the prices of both oil and gas continue to fluctuate. Fossil fuel and renewable energy prices, social and environmental costs are moving in opposite directions and it has become clear that future development in the energy sector will be predominantly in the new regime of renewable energy and to a certain degree natural gas-based systems and not the conventional oil and coal based sources (Herzog et al., 2001). The ability in producing smarter, lower-consumption energy based on renewable sources such as hydropower, solar, wind, biomass and biofuels, geothermal, and ocean and tides signifies an important step forward towards a sustainable energy future and creates new and exciting opportunities for cities (Khansari et al., 2014).

Despite the many benefits such as sustainability which alternative energy technologies such as wind and solar provide, renewable energy sources do however have negative consequences also (e.g. the energy and materials required for the construction of wind turbines or solar panels) (Boudreau et al., 2008).

THE SMART CITY

While the actual phrase Smart Cities is however not new itself as its actual origins may have originated from the Smart Growth movement of the late 1990s which promoted new policies for urban planning. Since 2005 several technology companies have adopted this phrase for the application of complex information systems to combine the operation of urban infrastructure and services including, buildings, water and electrical distribution, transportation and the provision of public safety. This phrase has since changed to include virtually any form of technology-based improvement in the planning, development and the operation of cities such as the deployment of services required for plug-in electric vehicles (Harrison & Donnelly, 2011). The general notion of the Smart City arose due to the utilisation of information technology for decision making by citizens, service providers and city government alike (Khansari et al., 2014).

With the current economic crisis which exists combined with the increasing expectations citizens have, cities experience increased pressure to behave in a smarter way providing both better and more efficient infrastructures and services, often at a reduced cost and it is this which has contributed to the term "Smart City" (Ballon et al., 2011).

The term or phrase "Smart City" is far-reaching and can incorporate a selection of different services which are offered to citizens and visitors alike and different ideas and objectives which vary from city to city despite being in the same country. However, the Smart City does offer one vision and that is resolving the troubling challenges by applying ICT to alleviate the impacts suffered due to rapid urbanisation and the associated follow- on effects (Steinert et al., 2011). The concept of the "Smart City" is relatively new in origin although it stems from or at least is a more advanced descendant of the older "information city", "digital city" and the "intelligent city" categories as recent years have witnessed the Smart City surpass its predecessors in popularity (De Jong, 2015). A controlled use of energy is a fundamental factor in the development of the Smart City concept as many of the ideas of the Smart City such as the charging of low polluting electric vehicles or the use of electric heat pumps as a different lower carbon dioxide emission central heating technology all require an energy source to function efficiently (https://www.gov.uk/government/ publications/smart-cities-background-paper). Despite its new origin, the Smart City concept still covers a wide cross-section of approaches and procedures to enrich the quality of urban living, the provision and the management of public services and long-term sustainability. After all a city's strength and standing depend on multiple factors including communication technology, waste management, public transport, access to clean drinking water, education, health and public safety (Barrionuevo et al., 2012). In an approach for tackling the many challenges encountered in urban management the concept of the "smart", "intelligent" or "cognitive" cities have all gained increasing attention as a method for dealing and resolving these challenges (Khansari et al., 2014).

Technology, particularly the use of ICT in connecting people, political institutions and business led to the birth of the Smart City. The Smart City's development is also attributed to the application of technology and other hi-tech solutions to urban problems as other technologies also play important parts as they aim to improve issues such as mobility and the environmental sustainability within the city (Dameri, 2013).

Smart cities are looked upon as recognition of the increasing importance which digital technologies play in both a competitive and sustainable future. The Smart City has awarded ICTs with the ultimate task of achieving urban development goals such as improving the quality of life for its citizens and creating sustainable growth. There have been six main areas identified where digital innovations should make a difference and they include: smart living, smart governance, smart economy, smart environment, smart people and smart mobility (Schuurman et al., 2012; Vanolo, 2014; Steinert et al., 2011).

The term Smart City is distinguished using the six conceptually distinct characteristics (Vanolo, 2014; Steinert et al., 2011; Lombardi, 2011):

1. **Smart Economy:** This involves the creation of business opportunities, providing citizens and businesses with broadband access and enabling networks to expand outside city centres. It also involves using electronic means in business processes of all kinds such as e-banking, e-shopping and e-auction.
2. **Smart Mobility:** This refers to local accessibility, the availability of ICTs and transport systems which are safe, modern and sustainable and the promotion of more efficient and intelligent transportation systems.
3. **Smart People:** Linked to the level of human qualification and social capital, originality, flexibility, tolerance, cosmopolitanism and the participation in public life.
4. **Smart Living:** This involves the quality of life which is imagined and measured in terms of availability of both cultural and education services, tourist attractions, social cohesion healthy environment, personal safety housing, access to a high-quality healthcare service.
5. **Smart Environment:** This is understood in terms of attractiveness of natural conditions, lack of pollution and sustainable management of resources and reduction in energy consumption through the application of novel technology innovations while promoting energy conservation.
6. **Smart Governance:** This relates to the decision-making process, transparency of governance systems, the availability of public services and quality of political strategies.

A Smart City indicates the usage of all available technology and resources in both a coordinated and intelligent manner to develop urban centres that are integrated, habitable and sustainable (Barrionuevo et al., 2012) who identified that the five types of capital which contribute towards a city's intelligence include the following:

- Economic (foreign investment, international transactions, gross domestic product (GDP) and sector strength).
- Human (talent, education creativity, innovation).
- Social (traditions, habits, families, religions).
- Environmental (energy policies, water and waste management, landscape).
- Institutional (civic engagement, elections and administrative authority).

The assumption of a smart city is that by having the correct information at the correct time, citizens, service providers, and city governments will similarly be enabled to make better decisions resulting in a better quality of life for the city's inhabitants and the overall sustainability of the city. However, information resulting from a Smart City application has a two-fold impact as firstly it changes the social behaviour of citizens towards a more sustainable utilisation of city resources (bottom-up) and secondly it enables service providers including utilities, service providers and city governments to provide a more sustainable and efficient service (top-down). All of the important infrastructures within smart cities such as roads, bridges, tunnels, rails, subways, airports, seaports, communication infrastructures, water power and major buildings are all monitored to maximise the services that are available to citizens including security services while optimising the use of resources. Real-time information regarding the status of urban services is required by cities so that they can improve public safety and offer sufficient infrastructure-based services including reliable electricity, safe drinking water, reliable communication and dependable transportation (Khansari et al., 2014).

Traditional cities however, differ from the Smart City as they are not able to enhance this provision of services simply due to the constantly changing conditions. The Smart City provides the required infrastructure for both citizens and officials enabling them to make more intelligent and informed decisions which plays an important role in dealing with challenges which relate to ecological, social, cultural and economic sustainability (Khansari et al.,2014).

The Smart City architecture involves the following three layers (Khansari et al.,2014):

1. **Human/Institutional Layer:** This layer comprises all residents, non-governmental organisations (NGOs), regulators and actors within the private sector who are involved in the creation of market dynamics. The addition of profit maximisation goals into the financial utility function of these social agents defines strategies at both the data network and the physical network levels.

2. **Data Layer:** This layer includes all the data gathering devices, information sensors local wireless and cellular networks which monitor the standing of various systems within the city. This layer also combines subsystems to make the overarching system ultimately more "smart."

3. **Physical Layer:** This layer comprises all the physical objects and infrastructures and their associated physical properties providing connectivity for the city's subsystems. It enables wireless sensors to be installed in components of the physical layer to collect monitored parameters and transfer this data to the data network layer and data network agents can likewise use those sensors/ actuators within the physical layer to monitor the performance of city systems and start control actions based on the economic optimisation scheme employed within the social network layer.

Smart cities ultimately can alter the environmental and social behaviours of citizens whether this involves the provision of information about mechanisms for reduction in energy consumption or updating on travel routes. They also enable smart governance and political participation among citizens and officials using ICTs such as e-governance and e-democracy. They control urban infrastructures such as systems of water, energy and land use, transportation supporting the use of renewable energy sources as a path to sustainable development. However, in adopting these technologies cites must deal with the challenges of privacy, security and government surveillance (Khansari et al., 2014).

A city is a Smart city when it operates in a sustainable and intelligent manner through the integration of all it infrastructures and services into one cohesive unit adopting the use of intelligent devices for monitoring and control ensuring both sustainability and efficiency. Sensors are a crucial element of any intelligent control system and a process is enhanced according to its environment and for a control system to be conscious of its environment it is fitted with a collection of sensors which it uses to collect data and using this data it can adjust its operations accordingly. One such example in the Smart City is the use of meters in determining gas, electricity and water consumption which would have been traditionally mechanical however, as smart metering implies a new technology of electricity meters have evolved from the manual procedure to automatic meter reading (AMR) reducing costs, improving accuracy of readings and leading to an advanced metering infrastructure (AMI) which unlike the (AMR) allows two-way communication with the meter driven by a growing understanding of the benefits of the two-way interactions between system operators, consumers and their loads and resources (Hancke & Hancke Jr, 2012).

The ability for Smart Cities to offer broadband connectivity to all city resident is limited by several factors which include the cost of deployment, operation, maintenance of the network. In a bid to overcome these challenges several approaches are adopted as unsurprisingly the largest challenge encountered is the financial cost of the smart cities broadband network. Several different approaches have been proposed by both national and regional government bodies and various industry stakeholders. These involve the sharing of some resources in a type of public-private partnership arrangement with various service providers. One primary approach proposed involves smart buildings and technologies assist to make the design, construction and operation of buildings both new and existing more efficient. ICT technologies and solutions include the building of management systems which run both heating and cooling systems which run accordingly to the needs of the occupant of the building and software which switches off personal computers and monitors in buildings when everybody has returned home. The building of energy management systems can reduce energy consumption by 5 to 40% and globally smart building technologies could eliminate 216 billion euros worth of carbon dioxide emissions (Steinert et al., 2011).

Essentially the formation of a Smart City is dependent on ubiquitous connectivity. Individuals, companies, governmental and non-governmental organisations, educational and healthcare institutions, public safety providers, objects such as buildings, sensors, all types of fixed and mobile devices and utilities and all the processes connected to a city require the ability to interact with each other in real time to enable the sharing of data and other content in a safe and secure manner. Ultimately the brain of the Smart City is the Smart Cities broadband network (Steinert et al., 2011).

DATA IN THE SMART CITY

The Internet integrated Smart City is becoming a reality within urban centres worldwide as this data-driven city is dependent on data collected from several different areas including buildings, infrastructures, people and third party data brokers. Exploration into how big data can be used for the building of smart cities in the developing and emerging regions of the world. Large technology companies such as Microsoft and Cisco assist with the design and security of the Internet of Everything which is fundamental to Smart Cities. Data-driven analytics delivers a wealth of information including the understanding of citizen and environmental relationships driving sustainable decisions and building urban survivability. Big data which describes the storage, collection, use and reuse of huge amounts of data collected from smartphones, the Internet, publicly and privately and increasingly from sensor devices of all shapes and forms. Big data is created in Smart Cities due to the use of technology and applications to both collect and analyse personal information from citizens and residents which is shared across functional areas and used in the urban planning process (Hiller & Blanke, 2016).

Due to the large volumes of data collected within a fully smart city an individual's every movement can be tracked and the data will show where he/she works, the different modes of transport they take, their shopping habits, places they visit and their proximity to other individuals. Such data will be centralised and easily accessible. Private companies will know more about people than they know about themselves (Hiller & Blanke, 2016).

SECURITY ISSUES

The large volume and quality of data collected will increase as the Smart Cities develops and as personal data is collected from smartphones, smart meters, plug-in electric vehicles and sensors making personal privacy a matter for concern (Bartoli et al., 2011)

Despite the many advantages which come from the collection of vast amounts of data interest regarding the subject of personal information and its privacy has grown raising serious issues. The widespread use of sensors and surveillance within the Smart City creates a society which ignores the privacy of individuals although big data is created in Smart Cities because of the use of the technology and applications to collect personal information from the city's residents for use in the urban planning process. For cities to become smarter, in many cases they will often have no other option but to collect either identifiable or personal information from its citizens or residents. Precise laws and regulations may be required for the Smart City, data collection, and the preservation of individual privacy and some areas have already taken and addressed these concerns. It has been proposed that both communities and individuals should have the option of making informed decisions with a complete list of who, why and whether their data is to be used and for what length of time this should take place (Hiller & Blanke, 2016).

Smart grids lead to a whole new set of challenges and a major component of future smart grids relies on both cyber security and control as the two most important challenges faced by the future smart grid are security and privacy while the two different classes of attack on the smart grid are either cyber or physical (Hiller & Blanke, 2016; Ardito et al., 2013; Saxena & Choi, 2015). These two different classes can be differentiated as the cyber-attack may cause a misbehaviour of physical components managed by software routines whereas the physical attack might lead to AMI bypassing to falsify accounting values or cause instability due to physical destruction (Ardito et al., 2013). Cyber-attacks can be divided in the following four categories (Hiller & Blanke, 2016):

1. **Device Attacks:** This is usually the first step of a complicated attack and it involves compromising the control of a grid device.
2. **Data Attacks:** The goal of a data attack is inserting, deleting or altering the data flow to get misbehaviours.
3. **Privacy Attacks:** This tries to use electricity usage data with the aim of learning or inferring the user's personal details/information.
4. **Network Availability Attacks:** The purpose of this type of attack is to overpower the communication and computational resources of the grid thus resulting in delay or failure of communication.

While the issues of security and privacy are, important issues faced by the smart grid and authentication is a key challenge in smart grid communication the modern power grid employs the use of supervisory control and data acquisition (SCADA) systems with communication protocols. However, these protocols used in these systems are often vulnerable to man in the middle (MITM) attacks, replay attacks, impersonation attacks and the cryptographic keys used in various devices of the system may become compromised as a result. The process of authentication involves proving the identity of a given system, including users, applications and devices as the typical smart grid network involves millions of devices all interacting with each other. Information exchange in the smart grid network requires all involved entities to be bi-directionally authenticated (Saxena & Choi, 2015).

DEFINITION OF THE SMART CITY

Although no agreed definition exists regarding a Smart City several components can be identified including: smart economy; smart mobility; smart environment; smart people; smart living; and smart governance (Vanolo, 2014; Steinert et al., 2011; Lombardi, 2011). Cities can be defined as smart when investment in both human and social capital together with investment in traditional (transport) and modern information and telecommunication infrastructure generates sustainable economic development and a high quality of life while promoting prudent management of natural resources (De Jong et al., 2015). Despite the lack of a widely recognised and accepted definition it is now however, possible to try and write the following comprehensive definition of the Smart City:

A smart city is a well-defined geographical area, in which high technologies such as ICT, logistic, energy production, and so on, cooperate to create benefits for citizens in terms of well-being, inclusion and participation, environmental quality, intelligent development; it is governed by a well-defined pool of subjects, able to state the rules and policy for the city government and development (Dameri, 2013).

SMART CITY PROTOTYPE

An early example or prototype of the Smart City is Santander a small port city in Spain. The city includes the installation of approximately 12,000 sensors placed in many different locations including under the asphalt, attached to street lamps and affixed to the top of city buses as well as many other locations. The sensors are designed to measure air pollution, locate the availability of parking spaces, have the availability to dim street lighting and even go as far as telling the garbage collectors when public bins are full. Street signs are fitted out with digital panels which display real-time parking information which is relayed back to a central control centre. City residents can download a selection of smartphone applications to obtain city information such as information on parking spaces, road closures, bus delays and pollen counts. Using this prototype, the city has saved about 25% on its electricity costs and 20% on garbage collection alone (Hiller & Blanke, 2016).

Several triggers can put cities on the road to becoming smart as the city may play host to a demonstration project which enables one or a few different companies to test their most innovative solutions and one such city is Songdo in South Korea were digital innovation projects are being tested (Aoun, 2013). This half-finished South Korean city of Songdo is a mixture of empty plots and gleaming towers built on landfill which was dumped into the Yellow Sea and when the 607 hectare Songdo IBD (International Business District) which comprises the heart of the city is complete many developers believe it will be the greenest, most wired city in the world (Strickland, 2011).

Business hubs like Songdo in South Korea unlike conventional cities are newly built and created from scratch. The city of Songdo was a joint undertaking by Cisco and real estate developers. Currently twelve years and $40 billion later it has resulted in a so-called "City of the Future" or "The World's Smartest City" and is home to roughly 70,000 people. The city of Songdo has completed roughly 60% of its proposed infrastructure and buildings with a current population of what is expected when the project is presumed to be completed in 2018. The city comprises numerous sensors monitoring everything from temperature to the flow of traffic. Environmental planning efforts consist of charging stations for electric cars and a water recycling system which will separate clean drinking water from water which is used to flush toilets. Household waste is disposed of directly from homes into underground tunnels where it is automatically treated and processed (Hiller & Blanke, 2016). The building of newly built hubs such as Songdo in South Korea and the design of such a compactly integrated system is moderately easy as there are fundamentally no limits outside of the usual financial constraints as to what urban planners can dream up and conversely the transforming of conventional cities presented much greater challenges (Barrionuevo et al., 2012).

These new cities have the distinct advantage of incorporating the vision of the Smart City from the outset and such cities are purposely placed or located in an area which is designed to attract businesses and residents which incorporates both ICT infrastructure and world class services. These cities often include broadband connectivity, renewable energy smart transportation and other smart city systems. Such cities are increasing in numbers worldwide especially in emerging markets including: Songdo IBD, South Korea; Meixi Lake, China; Masdar City, Abu Dhabi; King Abdullah Economic City, Saudi Arabia; Lavasa, India; and the newly announced Skolkovo, Russia. While some of these new cities have government sponsors others are however privately owned (Bélissent, 2010; Washburn et al., 2009).

THE FUTURE

The Global electrical grids are approaching the greatest technological transformation since electricity was introduced into the home. The Smart Grid is replacing the out-dated infrastructure which delivers power to both homes and businesses. The grid is a modernisation of the current electrical system and it enhances the ability of customers and utilities ability to monitor control and predict their energy usage (McDaniel & McLaughlin, 2009). Many benefits have already been derived from the use of analysing big data and they include areas such as healthcare where analysing huge datasets helped in determining drug interactions, negative side effects and the advantages certain drug therapies provide and in cities the big data within the smart grid enables electricity suppliers to have better control and monitor the usage of power within the city (Hiller & Blanke, 2016).

The smart electricity grid has also opened the doors to new applications with wide ranging influences which provides the ability to safely integrate more renewable energy sources (RES), electric vehicles and a more efficient and reliable delivery of power and the grid's ability for the prevention or restoration of outages (Giordano et al., 2011).

Smart cities are the future as both climate change and population movements demand them and 2015 saw the White House declare that $160 million in financial investment was set aside for further research in this area. India itself has a goal of developing a total of 100 smart cities as it anticipates that 50% of its citizens will reside in cities by 2050 in comparison to the 32% currently living there (Hiller & Blanke, 2016).

CONCLUSION

With the Global population continuing to increase and over half of the World's population selecting only urbanisation and residing in cities as opposed to rural areas, the resources of cities as a result are subsequently stretched due to the large numbers which select city life. With the growing numbers in cities there is consequently large volumes of data generated known as 'Big Data' which is created in large quantities, high velocity and in many different formats and stored for reuse. Many different forms of data are collected regarding citizens within cities including data concerning their use of electricity, gas and water usage of residents within a city are all collected.

Cities as a rule are becoming smarter and using smart grid analytics, the basic building block for the Smart City implementation they can deliver a vast amount of information about how citizens use the city's resources enabling the city government to make smart and more sustainable decisions ultimately leading to the city's sustainability. In the world's more developed countries the focus is more on the regeneration of urban areas as opposed to the development of new cities (Kuyper, 2016) and the smart city's popularity in Europe is based on several factors including the availability of sufficient funding (Vanolo, 2014).

The smart grid does not replace the existing electrical system as such a process would be impossible for both technical and economic reasons but it enables building on the current existing infrastructure while increasing the employment of existing assets and implementing both new services and features to permit the implementation of the new functionality (Gharavi & Ghafurian, 2011; Ardito et al., 2013).

REFERENCES

Al Nuaimi, E., Al Neyadi, H., Mohamed, N., & Al-Jaroodi, J. (2015). Applications of big data to smart cities. *Journal of Internet Services and Applications*, *6*(1), 25. doi:10.118613174-015-0041-5

Alawadhi, S., Aldama-Nalda, A., Chourabi, H., Gil-Garcia, J. R., Leung, S., Mellouli, S., . . . Walker, S. (2012). September. Building understanding of smart city initiatives. In *International Conference on Electronic Government* (pp. 40-53). Springer Berlin Heidelberg. 10.1007/978-3-642-33489-4_4

Albino, V., Berardi, U., & Dangelico, R. M. (2015). Smart cities: Definitions, dimensions, performance, and initiatives. *Journal of Urban Technology*, *22*(1), 3–21. doi:10.1080/10630732.2014.942092

Anthopoulos, L., & Vakali, A. (2012). Urban planning and smart cities: Interrelations and reciprocities. The Future Internet, 178-189.

Aoun, C., (2013). *The smart city cornerstone: Urban efficiency*. Schneider Electric.

Ardito, L., Procaccianti, G., Menga, G., & Morisio, M. (2013). Smart grid technologies in Europe: An overview. *Energies*, *6*(1), 251–281. doi:10.3390/en6010251

Ballon, P., Glidden, J., Kranas, P., Menychtas, A., Ruston, S., & Van Der Graaf, S. (2011), October. Is there a need for a cloud platform for European smart cities? In *eChallenges e-2011 Conference Proceedings, IIMC International Information Management Corporation* (pp. 1-7). Academic Press.

Bari, A., Jiang, J., Saad, W., & Jaekel, A. (2014). Challenges in the smart grid applications: An overview. *International Journal of Distributed Sensor Networks*, *10*(2), 974682. doi:10.1155/2014/974682

Barrionuevo, J. M., Berrone, P., & Ricart, J. E. (2012). Smart cities, sustainable progress. *IESE Insight*, *14*(14), 50–57. doi:10.15581/002.ART-2152

Bartoli, A., Hernández-Serrano, J., Soriano, M., Dohler, M., Kountouris, A., & Barthel, D. (2011). Security and privacy in your smart city. In *Proceedings of the Barcelona smart cities congress* (pp. 1-6). Academic Press.

Bélissent, J., (2010). *Getting clever about smart cities: New opportunities require new business models*. Academic Press.

Boudreau, M. C., Chen, A., & Huber, M. (2008). Green IS: Building sustainable business practices. Information systems: A global text, 1-17.

Cardone, G., Foschini, L., Bellavista, P., Corradi, A., Borcea, C., Talasila, M., & Curtmola, R. (2013). Fostering participaction in smart cities: A geo-social crowdsensing platform. *IEEE Communications Magazine*, *51*(6), 112–119. doi:10.1109/MCOM.2013.6525603

Carli, R., Dotoli, M., Pellegrino, R., & Ranieri, L. (2013). Measuring and managing the smartness of cities: A framework for classifying performance indicators. In *Systems, Man, and Cybernetics (SMC), 2013 IEEE International Conference on* (pp. 1288-1293). IEEE.

Cerrudo, C. (2015). *An emerging us (and world) threat: Cities wide open to cyber-attacks*. Securing Smart Cities.

Collier, S. E. (2010). Ten steps to a smarter grid. *IEEE Industry Applications Magazine*, *16*(2), 62–68. doi:10.1109/MIAS.2009.935500

Dameri, R. P. (2013). Searching for smart city definition: A comprehensive proposal. *International Journal of Computers and Technology*, *11*(5), 2544–2551. doi:10.24297/ijct.v11i5.1142

De Jong, M., Joss, S., Schraven, D., Zhan, C., & Weijnen, M. (2015). Sustainable–smart–resilient–low carbon–eco–knowledge cities; making sense of a multitude of concepts promoting sustainable urbanization. *Journal of Cleaner Production*, *109*, 25–38. doi:10.1016/j.jclepro.2015.02.004

Dohler, M., Vilajosana, I., Vilajosana, X., & Llosa, J. (2011), December. Smart cities: An action plan. In *Proc. Barcelona Smart Cities Congress* (pp. 1-6). Academic Press.

Domingo, A., Bellalta, B., Palacin, M., Oliver, M., & Almirall, E. (2013). Public open sensor data: Revolutionizing smart cities. *IEEE Technology and Society Magazine*, *32*(4), 50–56. doi:10.1109/MTS.2013.2286421

Fang, X., Misra, S., Xue, G., & Yang, D. (2012). Smart grid—The new and improved power grid: A survey. *IEEE Communications Surveys and Tutorials*, *14*(4), 944–980. doi:10.1109/SURV.2011.101911.00087

Forfás. (2013). *Priority Area K Smart Grids and Smart Cities Action Plan July 2013*. Academic Press.

Gharavi, H., & Ghafurian, R. (Eds.). (2011). Smart grid: The electric energy system of the future. Academic Press.

Giordano, V., Gangale, F., Fulli, G., Jiménez, M. S., Onyeji, I., Colta, A., ... Maschio, I. (2011). *Smart Grid projects in Europe: lessons learned and current developments*. JRC Reference Reports, Publications Office of the European Union.

Gungor, V. C., Lu, B., & Hancke, G. P. (2010). Opportunities and challenges of wireless sensor networks in smart grid. *IEEE Transactions on Industrial Electronics*, *57*(10), 3557–3564. doi:10.1109/TIE.2009.2039455

Gungor, V. C., Sahin, D., Kocak, T., Ergut, S., Buccella, C., Cecati, C., & Hancke, G. P. (2011). Smart grid technologies: Communication technologies and standards. *IEEE Transactions on Industrial Informatics*, *7*(4), 529–539. doi:10.1109/TII.2011.2166794

Hall, P. (1988). *Cities of tomorrow*. Blackwell Publishers.

Hancke, G. P., & Hancke, G. P. Jr. (2012). The role of advanced sensing in smart cities. *Sensors (Basel)*, *13*(1), 393–425. doi:10.3390130100393 PMID:23271603

Harrison, C., & Donnelly, I. A. (2011). A theory of smart cities. In *Proceedings of the 55th Annual Meeting of the ISSS-2011* (*Vol. 55*, No. 1). Academic Press.

Harrison, C., Eckman, B., Hamilton, R., Hartswick, P., Kalagnanam, J., Paraszczak, J., & Williams, P. (2010). Foundations for smarter cities. *IBM Journal of Research and Development*, *54*(4), 1–16. doi:10.1147/JRD.2010.2048257

Hernández, L., Baladrón, C., Aguiar, J. M., Calavia, L., Carro, B., Sánchez-Esguevillas, A., ... Gómez, J. (2012). A study of the relationship between weather variables and electric power demand inside a smart grid/smart world framework. *Sensors (Basel)*, *12*(9), 11571–11591. doi:10.3390120911571

Herzog, A. V., Lipman, T. E., & Kammen, D. M. (2001). Renewable energy sources. In Encyclopedia of Life Support Systems (EOLSS). Forerunner Volume-'Perspectives and Overview of Life Support Systems and Sustainable Development. Academic Press.

Hiller, J. S., & Blanke, J. M. (2016). *Smart Cities*. Big Data, and the Resilience of Privacy.

Kennedy, C., Steinberger, J., Gasson, B., Hansen, Y., Hillman, T., Havranek, M., ... Mendez, G.V., (2009). *Greenhouse gas emissions from global cities*. Academic Press.

Khansari, N., Mostashari, A., & Mansouri, M. (2014). Impacting sustainable behavior and planning in smart city. *International Journal of Sustainable Land Use and Urban Planning*, *1*(2). doi:10.24102/ijslup.v1i2.365

Kim, J., Filali, F., & Ko, Y. B. (2015). Trends and potentials of the smart grid infrastructure: From ICT, sub-system to SDN-enabled smart grid architecture. *Applied Sciences*, *5*(4), 706–727. doi:10.3390/app5040706

Kranz, J., Kolbe, L. M., Koo, C., & Boudreau, M. C. (2015). Smart energy: Where do we stand and where should we go? *Electronic Markets*, *25*(1), 7–16. doi:10.100712525-015-0180-3

Kuyper, T.S.T. (2016). *Smart City Strategy & Upscaling: Comparing Barcelona and Amsterdam*. Academic Press.

Li, H., Gong, S., Lai, L., Han, Z., Qiu, R. C., & Yang, D. (2012). Efficient and secure wireless communications for advanced metering infrastructure in smart grids. *IEEE Transactions on Smart Grid*, *3*(3), 1540–1551. doi:10.1109/TSG.2012.2203156

Ling, A. P. A., Kokichi, S., & Masao, M. (2012). *The Japanese smart grid initiatives, investments, and collaborations*. arXiv preprint arXiv:1208.5394

Lombardi, P. (2011). New challenges in the evaluation of Smart Cities. *Network Industries Quarterly*, *13*(3), 8–10.

McDaniel, P., & McLaughlin, S. (2009). Security and privacy challenges in the smart grid. *IEEE Security and Privacy*, *7*(3), 75–77. doi:10.1109/MSP.2009.76

Metke, A. R., & Ekl, R. L. (2010). Security technology for smart grid networks. *IEEE Transactions on Smart Grid*, *1*(1), 99–107. doi:10.1109/TSG.2010.2046347

Mohd, A., Ortjohann, E., Schmelter, A., Hamsic, N., & Morton, D. (2008). Challenges in integrating distributed energy storage systems into future smart grid. In *Industrial Electronics, 2008. ISIE 2008. IEEE International Symposium on* (pp. 1627-1632). IEEE.

Momoh, J. A. (2009). Smart grid design for efficient and flexible power networks operation and control. In Power Systems Conference and Exposition, 2009. PSCE'09. IEEE/PES (pp. 1-8). IEEE.

Morvaj, B., Lugaric, L., & Krajcar, S. (2011). Demonstrating smart buildings and smart grid features in a smart energy city. In *Energetics (IYCE), Proceedings of the 2011 3rd International Youth Conference on* (pp. 1-8). IEEE.

Moslehi, K., & Kumar, R. (2010). A reliability perspective of the smart grid. *IEEE Transactions on Smart Grid*, *1*(1), 57–64. doi:10.1109/TSG.2010.2046346

Mwasilu, F., Justo, J. J., Kim, E. K., Do, T. D., & Jung, J. W. (2014). Electric vehicles and smart grid interaction: A review on vehicle to grid and renewable energy sources integration. *Renewable & Sustainable Energy Reviews*, *34*, 501–516. doi:10.1016/j.rser.2014.03.031

Pirisi, A., Grimaccia, F., Mussetta, M., & Zich, R. E. (2012). Novel speed bumps design and optimization for vehicles' energy recovery in smart cities. *Energies*, *5*(11), 4624–4642. doi:10.3390/en5114624

Rodríguez-Molina, J., Martínez-Núñez, M., Martínez, J. F., & Pérez-Aguiar, W. (2014). Business models in the smart grid: Challenges, opportunities and proposals for prosumer profitability. *Energies*, *7*(9), 6142–6171. doi:10.3390/en7096142

Roncero, J. R. (2008). Integration is key to smart grid management. In *Smart Grids for Distribution, 2008. IET-CIRED. CIRED Seminar* (pp. 1-4). IET.

Saxena, N., & Choi, B. J. (2015). State of the art authentication, access control, and secure integration in smart grid. *Energies*, *8*(10), 11883–11915. doi:10.3390/en81011883

Schuurman, D., Baccarne, B., De Marez, L., & Mechant, P. (2012). Smart ideas for smart cities: Investigating crowdsourcing for generating and selecting ideas for ICT innovation in a city context. *Journal of Theoretical and Applied Electronic Commerce Research*, *7*(3), 49–62. doi:10.4067/S0718-18762012000300006

Steinert, K., Marom, R., Richard, P., & Veiga, G., & Witters, L. (2011). Making cities smart and sustainable. *The Global Innovation Index*, *2011*, 87–95.

Strickland, E. (2011). Cisco bets on South Korean smart city. *IEEE Spectrum*, *48*(8), 11–12. doi:10.1109/MSPEC.2011.5960147

Vader, N. V., & Bhadang, M. V. (2013). System integration: Smart grid with renewable energy. *Renewable Resources Journal*, *1*, 1–13.

Vanolo, A. (2014). Smartmentality: The smart city as disciplinary strategy. *Urban Studies (Edinburgh, Scotland)*, *51*(5), 883–898. doi:10.1177/0042098013494427

Washburn, D., Sindhu, U., Balaouras, S., Dines, R. A., Hayes, N., & Nelson, L. E. (2009). Helping CIOs understand "smart city" initiatives. *Growth*, *17*(2), 1–17.

Yu, Y., Yang, J., & Chen, B. (2012). The smart grids in China—A review. *Energies*, *5*(5), 1321–1338. doi:10.3390/en5051321

KEY TERMS AND DEFINITIONS

Biomass: Biomass is fuel developed from organic materials, a renewable and sustainable source of energy used to create electricity or other forms of power. It is renewable not only because the energy in it comes from the sun, but also because it can re-grow over a relatively short time period in comparison to fossil fuels which take hundreds of millions of years take to form.

Data: In the computing world data is information which is converted into a form which is usable or can be processed. In computing data takes the form of binary digital form whereas raw data is data in its most basic format.

Fossil Fuels: These take the form of coal, oil and natural gases which cause carbon dioxide emissions upon burning causing greenhouse gases (GHGs) and are the main cause of climate change.

Global Warming: This refers to a gradual increase in the temperature over the Earth's atmosphere usually caused by the greenhouse effect caused because of raised carbon dioxide levels and other pollutants.

Greenhouse Gases (GHGs): These include gases which add to the greenhouse effect by absorbing infrared radiation (IR). Carbon dioxide is just one such example of a greenhouse gas.

Information and Communication Technology (ICT): Information communication technology is information technology (IT) with the communication role added via telephones or wireless signals and computers also which allows for information and audio/visual signals to be transmitted, accessed, stored and used in some format.

Pollution: The introduction of a substance or solution into the environment which has either harmful or poisonous effects on the environment.

Sustainability: The environmental quality of not harming the environment or depleting natural resources to the point where they are no longer available thus still able to support the local community.

Urbanization: This refers to the percentage of people which live in urban areas compared to rural areas. Urban areas refer to built-up areas such as cities and towns whereas rural areas refer to areas which are not industrialized such as the countryside.

Utilities: This encompasses a broad range of services such as gas, water, electricity, waste removal, and sewage systems, and at times access to computing facilities such as the internet which are provided within households and businesses alike.

Chapter 5
Analysis of the Nexus Between Smart Grid, Sustainable Energy Consumption, and the Smart City

Luke Amadi
University of Port Harcourt, Nigeria

Prince I. Igwe
University of Port Harcourt, Nigeria

ABSTRACT

Since the 1990s, the field of smart grid has attempted to remedy some of the core development deficiencies associated with power supply in the smart city. While it seemingly succeeds in provision of electricity, it fails to fully resolve the difficulties associated with sustainable energy consumption. This suggests that the future of smart grid analytics in the smart city largely depends on efficiency in energy consumption which integrates sustainability in the overall energy use. This chapter analyzes the nexus between smart grid, sustainable energy consumption, and the smart city.

INTRODUCTION

Cities are the hub of economic development where energy consumption plays a key role. According to the United Nations Conference on Sustainable Development in Rio de Janeiro in 2012, half of humanity lives in cities. The urban population has increased from 750 million people in 1950 to 3,600 million in 2011. It is estimated

DOI: 10.4018/978-1-5225-3996-4.ch005

Copyright © 2018, IGI Global. Copying or distributing in print or electronic forms without written permission of IGI Global is prohibited.

that by 2030 almost 60% of the population of the world will reside in urban areas. The question of green energy and technology has not been adequately resolved in both developed and developing societies.

The strategies to have electricity consumption under efficient control are the basis of smart grid. This involves the control of energy usage based on the process data. The argument is that the process of achieving consensus on the future vision for technology development should lead to the mobilization of different environments around the idea of propagating a smart grid for a smart city. According to the European 2020 Strategy Document (European Commission, 2010b), Europe is suffering a period of structural transformation in the socio-economic framework. The priorities to favor the new model that Europe needs include institutionalization of a smart city dynamics. This includes a variety of activities and lifestyles.

In the United States, the first official effort at defining Smart Grid was provided in the Energy Independence and Security Act of 2007 (EISA-2007). The US Congress gave its approval in January 2007 and it was signed to law by President George W. Bush in December 2007. The EISA (2007) argued that a smart grid is an electrical grid which includes a variety of operational and energy measures including smart meters, smart appliances, renewable energy resources, and energy efficient resources.

Beyond secure and equitable energy consumption among the high-income societies, are the challenges of greenhouse gas emissions which are produced in cities. These appear to be at variance with the smart city prognosis. Equally, a number of studies explore "inclusive grid" within the sustainability contexts. This largely explores the persistent dichotomy between the rich and poor societies. Across Africa, parts of Asia and Latin America power outage points to opposite direction with the smart grid analytics.

This suggests the saliency of engagement with the question of smart grid analytics within sustainability contexts. In particular, this encompasses building solutions with the evolving needs and nature of citizens to accommodate adaptable foundations to shape the future of cities. This is premised on the notion to accommodate a number of evolving interests in the smart city dynamics including clean and green energy consumption, inclusive grid, grid control models etc. Thus, smart grid analytics emerged as basic building block to implement smart city.

Smart cities which could act as catalyzers of the foreseen energy, environmental and social equity are expected to play key roles in order to achieve a sustainable energy consumption. Energy issues relating to sustainable development were discussed at inter-governmental level for the first time at the 9th Session of the Commission for Sustainable Development (CSD-9), held in April 2001. Countries agreed that stronger emphasis should be placed on the development, implementation and transfer of cleaner, more efficient technologies and that urgent action is required to further develop and expand the role of alternative energy sources.

Sustainable consumption debate contends that consumption is a key lever to achieving more sustainable development, on the contrary, that unsustainable consumption patterns are major causes of global environmental deterioration, including the overexploitation of renewable resources and the use of non-renewable resources with their associated environmental impacts (European Environment Agency,2005).

Similarly, technological innovations have reduced the energy and material intensity of most products. However, the increasing volumes of consumed goods have outweighed these gains: Household energy consumption contributes to almost 30% to the total final energy consumption and is, after transport, the second most rapidly growing area of energy use (EEA,2005). Equally, industrial energy consumption also accounts for over 78% of total energy consumption(Barr et al., 2005).

The dynamic and rapid growth of cities will lead to scholarly and research trajectories on the future of smart grid. In particular, cities and urban planners must device new modalities in a sustainable manner that transcends 'meeting the energy needs of the present and future generation' to 'protecting the natural ecosystem where human and non-human species inhabit' .This stand point provides a new set of scholarship that proposes institutionalization of Marxian political ecology, a sustainable development initiative which matches political power and technological resource extraction with ecological equity and justice.

Against this background, this chapter is a foresight research which attempts to resolve the question regarding how smart grid analytics could help smart cities in sustainable consumption. The chapter proposes equitable mobilization of resources with emphasis on sustainability. In a distinct manner, it demonstrates that although increase in green energy producers and consumers may lead to more agencies with ability to provide smart grid analytics. However, that the concomitant challenge of 'efficient energy consumption' points out the saliency of sustainable consumption.

The assumption is that smart grid would be connected with process analysis that could mitigate deleterious energy consumption. To advance this knowledge, it is important to understand how process analysis in energy usage is manifested and how it can be enhanced at various levels in smart cities (Both at the micro and macro levels). The objective of this chapter is to determine a sustainable model for energy consumption. It posits that unsustainable energy consumption will pose more challenges to the future development and advancement of smart grid and communication systems, which may have adverse effects on smart grid analytics for the smart city. An alternative way of thinking about smart grid is explored one which prioritizes dynamics of sustainable and efficient energy consumption.

The chapter puts forward a novel knowledge on Marxian political ecology which centres on sustainability. It argues that a city cannot be smart when its key structures and components are unsustainable. Emphasis is laid on energy resource efficiency, equality, smart and green energy consumption which the chapter argues must be institutionalized in order for a city to lay claims to "Smart".

The rest of the chapter is structured as follows; theoretical framework, Conceptual issues: smart grid, smart city and sustainable energy consumption, smart grid and smart city: analysis of sustainable energy consumption, recommendation, future research directions and conclusion.

THEORETICAL FRAMEWORK

Several theoretical explanations have been provided on studies in smart grid analytics and the smart city in the broad field of sustainable development, reflecting on their key assumptions, analyzing key trends in their trajectories and suggesting future research agenda that might advance the divergent social trends, changes and theoretical threads in the field of inquiry (Castells,2000; Schor,2001; Shove,2003; Hamilton, et al; 2010; Galli, et al; 2011) . These spate of scholarly-reflections, have set up new research agenda that prove useful in understanding original impetus of sustainable development within various research threads. Sustainable energy consumption is one of such useful themes which seek to explore the question of the relation of present and future energy consumption in the smart city, within the smart grid analytics.

While endorsing the general theoretical orientation of Marxian political ecology, its less adoption in the contemporary real-world contexts driven largely by asymmetrical capitalist resource extraction has overly undermined a sustainable and equitable energy consumption. This has serious consequences for Smart grid analytics and the smart city. Thus, a distinctive component of the ecological critique of smart grid analytics and smart city debate is largely their exclusion of environmental concerns - both the understanding of how smart grid provides energy and the institutional core regarding the "clean or green" concerns within the production and consumption of energy. These less lucid concerns form the theoretical basis of the Marxian political ecology.

Ecology is the study of the relationships of organisms with one another and the relationship of organisms to their environment. Ecological Marxism is a neo Marxist theory which studies the primacy of capitalist power and inequality in access and use of natural resources (Speake & Gismondi, 2005:55; Harvey, 2005; Alkon& Agyeman, 2011;Amadi & Igwe,2016). Ecological Marxists draw attention to political and ecological factors in clarifying how material power and resource accumulation (e.g. capital, wealth, military power) mediate human society and nature relations (Biersack, 2006: 3; Bryant, 1998: 80).

Following the assumptions of the ecological Marxism, the chapter presents an overview and critique of the smart grid analytics and the smart city model. And then sets out in more details the ways in which the field of political ecology could fruitfully engage with the question of unequal access, underutilization and disproportionate energy consumption which challenges the smart city model. Thus, inequality remains a key strand of unsustainable consumption. Juliet Schor (2001) argued that "economic globalization, militarization, corruption, the monopolization of environmental resources, and the legacies of colonialism have meant that the global "South" doesn't consume enough—at least not in terms of basics such as food, clothing, shelter". Schor (2001) observed that "in 1999, per capita GDP in the Less Developed Countries (or "global South") was $3,410 (measured as purchasing power parity). By contrast the Developed Countries ("global North") enjoyed average per capita GDP of $24,430, a gap of about eight times. One third of the population in the global South (1.2 billion) lives on less than $1 per day; 2.8 billion live on less than $2 per day. Together, this 4 billion comprises two-thirds of the world's population".

Elizabeth Shove (2003,p.1) re-echoes an insightful concern and argued that it is evident that lifestyles, particularly in " the West, will have to change if there is to be any chance of averting the long-term consequences of resource depletion, global warming, the loss of biodiversity, the production of waste or the pollution and destruction of valued 'natural' environments".Schor(2001:4)further argued that; "The number of vehicles per person has increased; as has the size and luxuriousness of those vehicles. The culture of excess has yielded $20,000 outdoor grills; $17,000 birthday parties for teen girls at FAO Schwarz; diamond studded bras from Victoria's Secret; a proliferation of Jaguars, Porsches and other luxury cars; status competitions in stone walls; professional quality appliances for people who are never home to cook; designer clothes for six-year olds; and bed sheets costing a thousand dollars apiece".

On this basis, Marxian political ecology provides a theoretical and conceptual model that interrogates both smart grid and smart city studies and offers an alternative to the mainstream debates on equitable and sustainable energy consumption.

CONCEPTUAL ISSUES: SMART GRID, SMART CITY AND SUSTAINABLE ENERGY CONSUMPTION

There is need for conceptual clarity on the issues of smart grid, smart city and sustainable energy consumption by exploring these concepts and offering a way of thinking about how they could be interrelated within the understanding of city life and sustainable development.

Smart grid is the 'third great revolution' of electricity (Mazza,2004) which represents a wide range of 'changes in its production, delivery and use' including concepts in the present and future research agenda linked to the challenges of electricity supply. The wide range of terms associated with smart grid analytics and absence of consensus on scholars on a universal definition suggests divergent terminological and conceptual complexities. This points out that a number of concepts have been used to examine smart grid analytics encompassing complex systems and networks. The smart Grid Energy resource Centre in the United States considered the power grid within the context of optimal control, ecology, human cognition, glassy dynamics, information theory, microphysics of clouds, and many others. Mulgan(1991)had suggested the relevance of control of networks within the new economies of communication. Geoffrey Werner-Allen et al.;(2006, p.1)propose the deploying of a wireless sensor network on an active volcano .They suggest that sensor-network application for volcanic data collection relies on triggered event detection and reliable data retrieval to meet bandwidth and data-quality demands.

Quite apart from specific and proven smart grid technologies in use as explicated, Smart grid remains an evolving field of inquiry involving a set of related technologies which points out the need for conceptual clarity. In a number of studies, smart grid represents an advance on electricity as it aims at energy consumption in relation to a variety of uses in the society. A number of research had recently been written in the field of smart grid, providing its salient remit, reviewing key trends in its content and possible future research prospects (Lichtenberg,2010; Hamilton et al., 2010; Galli et al., 2011). The concept of power line communication, which examines the use of electricity infrastructure for data transmission, is equally gaining scholarly relevance in smart grid analytics (Berger et al.;2013).

Since the 1990s and early 2000s, there have been proliferation of grid programmes in most technologically advanced societies of the world. This includes the Telegestore system developed in Italy in 2005, the Austin, Texas mesh network which dates around 2003, and the Boulder, Colorado smart grid in which was set up in 2008. Other grid programmes in the United states include *IntelliGrid* – a brainchild of the Electric Power Research Institute (EPRI), *Grid 2030* developed by the U.S. electric utility industry and the *Modern Grid Initiative (MGI)* among others. The Pacific Northwest National Laboratory (2007) report shows that wide area networks were revolutionized following the Bonneville Power Administration (BPA) which broadened its smart grid research with prototype sensors simulated for rapid analysis of anomalies in electricity quality over very large geographic areas. This resulted in the foremost operational Wide Area Measurement System (WAMS) in 2000. A number of countries gradually assimilated the model. For instance, China reopened a detailed national WAMS system on completion of its 5-year economic plan in 2012 (Yang,2001).

Scant conceptual exploration on the smart grid model underpins theoretical and conceptual impasse which gives rise to various analytical difficulties in arriving at the extant terminological underpinnings of the concept. For example, institutional exploration of smart grid underscores the corporate dynamics of consumption (Fine & Leopold,1993), while micro grid examines individual or household network consumption patterns (Brohmann, et al;2013). Gender dynamics of grid consumption have also been given some conceptual attention (Thompson,1996). There are debates on grid integration for renewable energy linking aspects of communication for resource efficiency (Yu,et al;2011).

The complex conceptual exploration provides the original impetus to investigate smart grid which the present chapter seeks to explore. This is on the basis that the question of the relation of smart grid to sustainable energy consumption and the smart city appears not to have been adequately conceptualized. A common thread in the conceptual issues raised on smart grid is the application of digital processing and communications to the power grid, making data flow and information management central to the smart grid. There is a widespread assumption that smart grid analytics have had a robust relevance within the sphere of technological advancement, particularly in the smart city.

A common thread in the conceptual issues raised on smart grid is the application of digital processing and communications to the power grid, making data flow and information management central to the smart grid. There is a widespread assumption that smart grid analytics have had a robust relevance within the sphere of technological advancement, particularly in the smart city.

The transition from "smart machine" (Zuboff,1988) to "smart city" has gained recent scholarly attention. This includes debates on the rise of the "network societies" (Castells,2000), "power and networks" thesis(Huges,1983). Research on networks examines the modern wireless networks which have been an important conceptual exploration of the smart grid studies. This encompasses issues of "network congestion" (Ahmad et al.; 2009). There are studies on technological innovations to salvage congestion including several proposed algorithms. Both the "global city" (Sassen,2005) and the "green or eco-city" architecture aim at sustainable environmental impact to mitigate unsustainable energy, water and food consumption, pollution and waste (Register, 1987; Lehmann,2010). Technological innovations linked to smart city have also been conceptualized as "invention of comfort"(Crowley. 2003; Shove, 2003). This suggests the ease and comfiture of technological advancement.

On its part, the term smart city has been widely used in a variety of scholarships including seminars, journal articles, conferences and in particular among social scientists, engineers, city planners and a wide range of professionals. This suggests a multidisciplinary discussion which accounts for the emergence of various technologies, such as Information communication technology(ICT), Social Networks, Web Applications and Internet Technologies, immersive technologies, virtual and augmented reality, wearable technologies, cloud computing, Data science, Big Data insights etc.

The internet of things(IoTs)have also provided useful insights on the smart city debates. According to the Digital Agenda for Europe (2014), the Smart City concept means smarter urban transport networks, upgraded water supply and waste disposal facilities, and more efficient ways to light and heat buildings. And it also encompasses a more interactive and responsive city administration, safer public spaces and meeting the needs of an ageing population. Smart City is a new paradigm for the integration of Internet Technology in the Urban context.

"The Future of Cities is Smart", and similar mantra have emerged to suggest the plausibility of the smart city paradigm including several partnerships as well as various research organizations worldwide which are currently engaged in Smart City potentials and applications. Advanced Internet Technologies currently provide smart applications for virtually every sphere of human endeavor.

Smart City internet technologies are deployed in contemporary times as the basis of scientific debate for the new architect of Urban Computing. Smart Cities technologies and applications have received growing attention in recent years from various perspectives. The thriving numbers behind their adoption and exploitation in different application contexts have captured the attention of Internet Technology specialists, computer engineering and business researchers that, in the past years, have been trying to decipher the phenomenon of urban computing, its relation to sustainable energy consumption and its implications for new research opportunities that effect innovations in modern living, way of conducting business and developing economy for sustainability.

There are concerns in the smart city schema related to dichotomy arising from globalization. Thomas Barnett (2004) had argued on the penetration of globalization and disconnection of the outside world. David Ronfeldt, links this to "cyberocracy" a concept he deployed to explain the new state as a proliferation of transnational network rooted in a primarily cybernetic vision of the government that rules by the use of information and of society. Ronfeldt (1992) connects this to the concept of the "nexus-state," to explain a new network state entity enabled primarily by technology and networks. There are works which justify technological advancement as a 'modernization paraphernalia' and panacea to ecological challenges popularly known as the ecological modernization school (Spaargaren,1997). The ecological modernization school has been critiqued by the ecological Marxian scholars as less inclusive and rather exploitative. They contend that relying on technology alone might fail to deliver inclusive development (Harvey,2005; Schor,2005).

Despite its robust offshoot there are conceptual reflections on sustainable energy consumption (Barr et al., 2005). This becomes perceptive as the mainstream notion of smart grid suggests a "smart and greener model'(Mazza,2004). This orientation reinforces the conceptual revaluation of sustainable energy consumption.

Energy consumption arguably since the post-industrial revolution era has been overly influenced by the logic of capitalist pattern of consumption (Harvey,2005). It is not enough to dismiss deleterious effects of unsustainable energy consumption in an attempt to forecast the future of smart cities in an increasingly capitalist social formation. Thus, unsustainable energy consumption has serious consequences for smart city. This is discernible from the sustainable development perspective. Conceptualized in the Brundtland Commission Report (1987)as development that meets the needs of the present generation without compromising the ability of the future generation from meeting theirs, Sustainable development has three key development threads namely social, economic and environmental. In this regard, the chapter conceptualizes aspects of the environmental component linked to sustainable energy consumption.

In the last thirty years or so, there has been a proliferation of studies on various aspects of sustainable development. However scant studies have linked sustainability to smart grid analytics and the smart city specifically in the context of energy consumption. This has resulted in disciplinary attention to sustainable energy consumption linked to the smart grid analytics. Sustainable consumption has become central in virtually every field of development inquiry, particularly the dynamics of capitalist energy resource appropriation or extraction and its effects on the wider society. This puts the smart grid analytics at the centre of the sustainability debate resulting in a number of new useful research agenda, which among others point out the need for sustainable energy consumption.

Conceptual advances in this direction have provided seminal arguments. For instance, Schor (2005) offers a useful insight on the effects of disproportionate consumption of the advanced societies notably the United States and the UK, while Daniel Milner (2012) suggests the modest consumption patterns of Norway. Such comparative analysis has been insightful in current debates on global unsustainable energy consumption. This presents a persuasive strand of the implications of unsustainable consumption. On this basis, sustainable energy consumption as argued aims at a conceptual model of smart grid that offers an alternative to the largely Eurocentric paradigm in smart grid analytics.

More inclusive studies on smart grid analytics are needed to address the "global digital divide" which not only affects smart grid and smart city policy scholarship rather undermines the future of sustainable and equitable energy consumption. Thus, the arguments for further studies will build on prior conceptualization of smart grid and smart city.

The sustainability theoretical debates create linkages between the overall energy consumption implications to both to human and non-human species and the ecosystem. Responsible consuming has been an important question in response to sustainability signals (Warde, 1999; Davidson & Hatt, 2005; Dobbyn & Thomas,

2005). This accounts for the changing patterns of energy consumption often ignored in policy and research agenda.

There has been evidence of disparity on the smart grid scholarship largely propagating the needs of the developed societies. For instance, Gontar et al.; (2013) argued that smart city concept is an 'attempt to answer the following problems: urbanization, aging of social infrastructure in developed countries, cutting CO^2 emissions'.

The nexus between smart grid and sustainable consumption model tends to require more inclusive approach encompassing the urbanization dynamics of both the developed and developing societies. Thus, concepts like "smart grid" and "smart city" form part of resurgent debates on urbanization which are part of everyday lives of humanity. Yet, they are also contentious, as several political, social and technological interactions are often made on the basis of specific interests or lifestyles which might be at variance with the tenets of sustainability.

Smart Grid and Smart City: Analysis of Sustainable Energy Consumption

The ecosystem is in a state of flux as it is impacted by a number of factors notably rapid urbanization, population growth, technological advancement and global connectivity. These external factors will have significant impact on smart grid analytics in the future. These suggest that the question of smart and sustainable city has never been adequately resolved.

Analysis should advance in a way which provides analytical and conceptual capacity to explore the interrelatedness and overlapping spheres of practices and processes within both concepts. Thus, while smart grid analytics is fundamental to a smart city, it should be conceptualized within sustainability strand, distinct from the capitalist resource exploitation. Shove (2003) provides such interface between "convergence and conventions of comfort", in the context of "cleanliness and convenience" regarding analysis of a "green consumption".

Another concern is the cost of institutionalizing a smart grid particularly among the poor societies. Recently, the World Economic Forum (2016) reported that a transformational investment of more than $7.6 trillion is needed over the next 25 years (or $300 billion per year) to modernize, expand, and decentralize the electricity infrastructure with technical innovation as key to the transformation.

Although we now live in an almost digital world, Mulgan, (1991)argues 'that the late twentieth century world is covered by a lattice of networks'. Castells (2000) described this as the 'network society' which will develop 'through synergy among relevant theorizing, computational literacy, and sociological imagination dominated by technology' and internet hypertext. Several technological systems still require inclusive and sustainable components.

As recent trends suggest, the rise in smart cities in the 21st century including smart grids for signaling and conditioning energy consumption including meters, sensors and actuators for gauging, point out the increasing need to explore the sustainability linkages. Current trends in global energy use are far from sustainable. Oil demand continues to grow, while experts expect a historic peak in oil production within the next 20 years. Carbon dioxide ($CO2$) emissions from fossil fuel combustion in 2002 were about 13% above the 1990 levels, whereas a stabilization of the climate would demand a reduction by 50%until 2050 and further reductions thereafter (UNEP, Centre on Sustainable Consumption and Production (CSCP),2005).

There is significant variation on energy consumption at individual levels across countries and regions of the world which has implication for sustainable consumption. Evidence of unsustainable energy consumption includes the rise in pollution, flaring of gas, ecological disruptions etc. For instance, while OECD energy consumption would need to decrease, there is evidence of limited access to energy supply in the low-income societies. The world's richest people, earning over 20,000 US dollars per annum, consume nearly 25 times as much energy per person as the poorest people (UNEP, CSCP, 2005). On the other hand, currently, nearly one third of the world population has no access to electricity and another third has only poor access (UNEP, CSCP, 2005). Reliance on traditional fuels for cooking and heating can have a serious impact on health and the environment. This has different implications and impacts in terms of ecological breakdown, supply voltage, power rating and usage. It is evident that there are different needs for energy consumption serving various social, economic and technological needs. This may yield different results positively or negatively as far as efficient and optimal use of energy is concerned.

Sustainable energy consumption suggests policy relevant dynamics aimed at eco efficiency. This reinforces the notion that sustainable energy consumption is aimed at meeting the energy needs of the present generation without tainting energy use for the future city users. It integrates energy into a single, smart-power analytics to meet energy consumption needs of the city. It equally allows multifarious -integration of energy consumption as an inclusive logic as well as a modality for efficient use. High voltage and complex systems could be made compatible through modalities of efficient and sustainable consumption. This includes stepping the grid down to the poor which forms the thesis of a research agenda that argues that both city and grid are smart to the extent that they are inclusive (gender, class, race), affordable (pro poor), clean and green (pro nature), value driven, renewable, replicable, cost effective (efficient).These are constituents of sustainable grid /sustainable city.

A grid is 'smart' to the extent that it is compatible, inclusive, and responsive and conforms to the tenets of efficient energy consumption. It links equity with energy consumption. This underpins the fact that there have been a number of challenges associated with the task of providing improved quality of energy services in modern cities and urban centres. This largely includes the transformation of the global energy system, creating more viable service experiences.

In particular, this will guide the future of smart grid analytics as it will show how sustainable energy consumption could be deployed in a wide range of application which justifies the classification of a city as smart or otherwise. This encompasses the broad fields of transportation, automotive systems, manufacturing/production, consumer electronics, communications etc.

Irrespective of its robust analytics, the dearth of studies exploring the revolutionary role of the Smart grid within sustainability contexts is less explored. This is a key requirement for the advancement of a future research direction for the smart city. Analyses of technologies for the integration and interconnection of citizens, businesses, government and the natural environment have been superficial. This does not match the actual importance of the proliferation of the smart city scholarship. The specific changes related to unsustainable energy consumption, will shape the future and nature of smart grid analytics over the coming decades. Thus, the changes(high)energy consumption largely industrial and commercial among the high-income societies will taint the sustainability dynamics of the smart city.

Against the background of these trends, a number of alternative and critical perspectives had emerged to interrogate the sustainability assumptions of the smart grid analytics and the smart city (Alkon & Agyeman, 2011). Sustainable smart city transcends deploying technologies and divergent applications, rather involves a more direct question that boarders on the preservation or conservation of energy use to the broader elucidation of the persistent challenges of transition from exclusive to inclusive micro-grids technology access, which identifies individual energy needs with specialized and equitable distribution model among end users.

Bridging rural/urban dichotomy, which creates "grid divide" accounts for a more robust grid. The re-engagement with this debate offers alternative inclusive model particularly among the poor societies of the Third World. Thus, as the developing societies are rapidly urbanizing, sustainable grid becomes a critical component of contemporary urbanization. This largely includes protecting the ecosystem from hazardous effects of energy consumption pointing out the need for grid accountability, reliability, responsible and transparent grid which ensures stability.

Scant research inquiry had examined sustainable smart city including aspects of eco efficiency, city ecological footprints, eco labeling and triple bottom line. Smart city research for sustainable innovation should be deepened in contemporary sustainable development studies as important bloc in contemporary urbanization. Eberle et al.., (2004) identified sustainable consumption as a more ecological but also socially ... way of buying and using goods and services.

The challenges of disproportionate consumption among the high, middle and low-income countries create disparity regarding access and equitable energy consumption. For instance, there are concerns on billing systems, smart meters and sustainability challenges. This includes the question regarding accountability, precision, remote control, censorship, variable rate pricing. Thus, grid preference could be given to some sections of the society, country or class of individuals than others. A socially inclusive grid accounts for sustainability.

Control mechanisms are central concern to the future of smart grid for a smart city. Who controls the grid? And what yardsticks make for a viable control mechanism? These include indices that account for meter reading dynamics.

The extent to which smart grid guarantees corporate and individual privacy is less clear. This suggests the overriding influence the grid providers might wield which could be at variance with sustainable production of energy and implicitly underscores supplier and end user disparity.

Equally, complex and divergent security challenges are linked to sustainable grid consumption. This points out the need for modalities to protect and mitigate threats arising from grid related hazards.

SOLUTIONS AND RECOMMENDATIONS

The proliferation of smart grid scholarship and absence of policy and research consensus on the future of smart city and smart grid has been a central concern of the sustainable grid scholarship. This has led researchers to advance knowledge on future trajectories of smart city within the sustainability paradigm.

Smart grid analytics should be designed within natural resource friendly distribution linked to complex but mutually related networks based on cost effective and eco- friendly processing power. Pro poor grid architecture should be an integral part of sustainable grid.

This includes affordable grid, reliable grid-monitoring dynamics such as meters, grid maintenance and management mechanisms. All point out the saliency of smart grid and need for a policy shift in conceptualization from the stand point of grid model to sustainable smart grid. This increasingly results in new modes of inquiry including debates on smart micro-grid architecture, smart city authentication and authorization model, metering and billing, power surge and economic power distribution models, smart city priority scheduling, grid surge and control models etc.

Cognizant of the central role cities play in the overall development of an economy, being the hub of social, technological and economic transactions, including innovation, hospitality, connectivity and creativity, as well as various types of service delivery centers, sustainable city development models becomes a priority area within the smart city studies, global energy efficiency action plan is inevitable. Proponents argue that energy efficiency is building bloc to sustainable future (Blok,2005; Young, 2008).

Technological advancement suggests that several external factors will have significant impact on smart grid analytics in the future. This will have effects on the sustainability component of smart grid which inevitably will impact cities. With the interconnectedness of the city and the economy, government, the people and social interactions, and the ecosystem- a complex interplay of challenges are imminent as increasing changes in consumption pattern and lifestyles of people will affect the level of energy consumption, and point to the need for improved smart grid models to contain rising demands on energy.

'An increase in green energy producers and consumers may lead to more agencies with potentials to provide smart grid analytics'. To achieve these objectives, there are a number of salient initiatives: regulatory dynamics on energy consumption, the institutionalization of green and smart energy, efficient use of resources, and social inclusiveness among others.

Given the importance of Smart Cities, it is mandatory to define and standardize indicators to measure the evolution of cities to the challenge of achieving higher competitive rates within the sustainable cities. This helps to critically explore, identify and align certain indicators as measurable indices to determine how "sustainable" a "smart city" is. The aim is to transcend rhetoric and move into realities and specifics. For instance, are there indicators measuring corruption or terrorism in a smart city? What are the evaluation models for determining the smart nature and variables of an alleged smart city? How has ecological challenge of deleterious resource extraction including capitalist exploitation impacted the natural environment? How has climate issues such as vulnerability of climate change and uncertainties of global warming and natural disasters notably tsunamis, cyclones, hurricanes, earth quakes etc been tailored for possible mitigation and mainstreaming into the smart city model? These and similar concerns are pertinent in smart city and sustainability studies.

Existing studies on the adoption of energy-efficient measures both industrially and for household purposes are examined from a multi-disciplinary conceptual lens. This includes geographers, sociologists, regional and town planning studies, consumer behavior studies as well as market related dynamics. Research and policy discourse on smart electricity metering should be enlarged to check consumption levels and preferences of various end users.

It comprises institutions, practices and forms of tariff, yet all of these practices are less revolutionary in charting a sustainable energy consumption as they may not ameliorate emissions and unequal access. Kannberg et al.;(2003) provide a number of benefits to smart grid as a modernized energy system. This among several others includes reduction in power consumption particularly at the consumer side conceptualized within the *demand side management*. Such amelioration of consumption facilitates including load balancing, mitigates power failures. It is argued that increased efficiency and reliability of the smart grid will save consumers some money and aid in reduction of CO_2 emissions.

We suggest that sustainable energy points out that energy-saving measures should include amelioration of waste, low-cost or no-cost measures to check both capital investment and individual consumption patterns. There is consensus among experts that the implementation of more sustainable consumption behavior requires not only awareness among consumers, but also changed social and economic structures: Consumption is a "socially constructed historically changing process" (Bocock, 1993: 45). Both UNEP (2005) and Centre on Sustainable Consumption and Production (CSCP)(2005) suggest the following measures to sustainable energy consumption;

1. Demand-side energy efficiency (also termed energy end-use efficiency): This most important option relates to technical, organizational and individual measures to reduce the final energy needed to heat/cool our houses, produce goods etc.
2. Co-/tri-generation: Introduction of on-site co- or tri-generation of heat, cold, and power can dramatically improve energy efficiency on the supply side. This option is largely related to the issues of energy generation and distribution. It is, therefore, only further discussed here in relation to its applications at consumers' sites (e.g. in industry or public buildings).
3. Renewable energy: The third option is renewable energy produced and used onsite through biomass or solar thermal collectors etc. as well as that fed into the electricity grids.
4. Limiting energy services: The final option could be to limit the amount of energy services we use (e.g. by capping dwelling floor space) to a level sufficient to cover our energy-related needs.

Similarly, adopting equitable energy access policies have been suggested. IEA (2009b) posits that, "sound statistical data…and a clear description of the (energy services) situation" are the first of the preconditions for successful rural energy access policies.

Again, the understanding of the smart grid analytics within a smart city should not be treated in isolation of the prevailing global asymmetry in which a colossal technological gap exists between societies of the affluent North and the poor South. Thus, how the smart city is framed to meet the demands and needs of the increasingly urbanizing poor countries of Asia, Latin America and Africa is less lucid. These are some of the salient issues of future development concern.

FUTURE RESEARCH DIRECTIONS

The future research agenda is confronted with a number of trends. A human centered grid should drive much of the smart grid analytics. Johnson (2009) had proposed a "human centered information integration for the smart grid".This has not been given adequate attention.

Samir Succar and Ralph Cavanagh (2012) have examined "the promise of smart grid" and came to a conclusion that the smart grid can give us cleaner air, better health, lower electricity bills, and reduced carbon dioxide (CO_2) emissions in the atmosphere. Beyond the alleged energy efficiency, future research agenda will be able to examine more critical and in-depth evaluation of these claims. Particularly in the context of "cost effective smart grid". The point has been that the grid analytics have been quite expensive.

Next-generation transmission and distribution infrastructure will be better able to handle possible directions of the valuation of more sustainable producers and consumers. Smart Grid which is easily the next big technological revolution since resurgence of internet has a critical role to play in the future cities.

From the foregoing, the future policy direction of a smart city one which is built on sustainability is imperative. There is need to review and determine a new vision on smart grid analytics and the smart city. The move toward smart city platforms will require sophisticated models and approaches for a sustainable and secure future. There is need for new strategies to maximally leverage on the new technologies in the daily life.

The likely trajectories of sustainable city proposed in this chapter is inclusive and participatory within all facets of city development including communication, transportation, urban spaces etc which are linked to the understanding of the linkages between smart grid analytics for a smart city. The point the chapter makes at intervals is that future research agenda should focus on the process and dynamics of achieving sustainable consensus on the future vision for technology development that could direct access and alignment with divergent environments within the idea of smart city in the neo liberal order.

Ideas linking smart city to neo-liberal resource consumption have been scant. This suggests the saliency of a research agenda that should keep track of resource use and resource conservation. Driven largely by the notion of capitalist exploitation there is increasing need to always interrogate the place of environmental sustainability in the smart city research. This includes issues such as green city, eco efficiency, urban renewal, triple bottom line, eco labeling, ecological footprint and smart city accounting (Amadi & Imohita, 2017). With the notion of the smart grid as a fundamental attribute of a third industrial revolution and neo-industrialization, future research should keep track of the feasibility of such attributes within the core tenets of sustainability blocs namely social, economic and environmental in a distinct manner to explore patterns of equitable and resourceful energy consumption. The basis of this trend is to deploy control mechanism for mitigation of deleterious energy use. Thus, a new research direction that emphasizes process mining, sustainability analysis, inclusive, pro nature, pro poor and pro energy consumption particularly among the poor societies of the global South will provide relevant future trajectories for all stakeholders.

In particular, future research direction should chart a collective cause aimed to understand and deploy sustainability strategies including the strategic role of smart grid analytics at different levels of the smart city, involving different stakeholders both domestic, industrial and end users.

CONCLUSION

The point this chapter has been making and continues to make is that increase in green energy production and consumption may lead to more agencies with capacity to provide future smart grid analytics. However, the concomitant challenge of 'efficient energy consumption' remains inevitable within a capitalist system riddled with inequality. This underscores the saliency of sustainable energy consumption policy response.

This chapter at intervals communicates and disseminates the urgency of sustainable energy consumption policy response. It insists that research agenda in smart grid analytics and sustainable smart city is essential. The aim is for efficiency in energy consumption and to preserve the smart city, promote- smart grid analytics and the user experience with the provision of sustainable services for all stakeholders including businesses, government and end users. There is ample projections on the state-of-the art potentials of Smart City Systems and applications and related innovative technologies. However, the sustainability model driving the smart grid process remains less lucid. This has been contestable worldwide and to a large extent shows how the future of smart grid could be analyzed and in particular deepened within the sustainability frontiers in the advanced societies and how the poor societies could be included in the smart grid analytics.

The chapter partly initiates a dialogue between the smart grid analytics and smart city on the basis that the smart city agenda has not been adequately and sufficiently explored within the broader sustainability context. Furthermore, this sustainability orientation provides a rationale for advancing an inclusive, renewable, and pro poor analytics. It is suggested and desirable to define a smart city within sustainability contexts constituted by resource efficiency. This will help address the question of the relation of capitalist resource consumption, the natural environment and the poor.

This model as argued moves smart city beyond the largely technological advancement frameworks that prioritizes capitalist resource exploitation, to an understanding of the natural resource as a set of inclusive and collective sphere for the future of the natural ecology, human and non-human species.

REFERENCES

Ahmad, S., Liu, M., & Wu, Y. (2009). *Congestion games with resource reuse and applications in spectrum sharing.* CoRR, abs/0910.4214

Alkon, A., & Agyeman, J. (2011). *Cultivating food justice: race, class, and sustainability.* Cambridge, MA: The MIT Press.

Amadi, L., & Igwe, P. (2016). Maximizing the Eco Tourism Potentials of the Wetland Regions through Sustainable Environmental Consumption: A Case of the Niger Delta, Nigeria. *The Journal of Social Sciences Research, 2*(1), 13–22.

Amadi, L., & Imoh-Ita, I. (2017). Intellectual capital and environmental sustainability measurement nexus: a review of the literature. *Int. J. Learning and Intellectual Capital, 14*(2), 154-176.

Barnett, T. (2004). *The Pentagon's New Map.* Putnam Publishing Group.

Barr, S., Gilg, A., & Ford, N. (2005). The household energy gap: Examining the divide between habitual- and purchase related conservation behaviors. *Energy Policy, 33*(11), 1425–1444. doi:10.1016/j.enpol.2003.12.016

Berger, L., Schwager, A., & Escudero-Garzás, J. (2013). Power Line Communications for Smart Grid Applications. *Journal of Electrical and Computer Engineering.* 10.1155/2013/712376

Biersack, A. (2006). Reimagining political ecology: culture/power/history/nature. In A. Biersack & J. B. Greenberg (Eds.), *Reimagining political ecology* (pp. 3–40). Durham, NC: Duke University Press. doi:10.1215/9780822388142-001

Blok, K. (2005). Improving Energy Efficiency by Five Percent and More per Year? *Journal of Industrial Ecology, 8*(4), 87–99. doi:10.1162/1088198043630478

Bocock, R. (1993). *Consumption.* London: Routledge. doi:10.4324/9780203313114

Bryant, R. (1998). Power, knowledge and political ecology in the third world: A review. *Progress in Physical Geography, 22*(1), 79–94. doi:10.1177/030913339802200104

Castells, M. (2000). *Toward a Sociology of the Network Society.* American Sociological Association.

Crowley, J. (2003). *The invention of comfort.* Baltimore, MD: Johns Hopkins University Press.

Davidson, D., & Hatt, K. (2005). *Consuming Sustainability Critical Social Analysis of Ecological Change.* Fernwood Publishing.

Digital Agenda for Europe. (2014). *The EU explained: Digital agenda for Europe European Commission European Commission Directorate-General for Communication Citizens information 1049.* Publications Office of the European Union

Dobbyn, J., & Thomas, G. (2005). *Seeing the light: the impact of micro-generation on our use of energy.* Sustainable Consumption Roundtable, London, UK.

Eberle, U., Brohmann, B., & Graulich, K. (2004). Sustainable consumption needs visions. Position Paper. Institute of Applied Ecology, Öko-Institut, Freiburg/ Darmstadt.

European Commission. (2005). Doing More with Less: Green Paper on energy efficiency. Brussels: Author.

European Commission. (2010b). *Communication From the Commission to the European Parliament, The European Council, The Council.* Brussels: The European Central Bank, The Economicand Social Committee and the Committee of The Regions.

European Environment Agency. (2005). The European environment — State and outlook 2005. Copenhagen: Author.

European Environment Agency (EEA). (2005). *Household consumption and the environment.* EEA Report No. 11/2005. Copenhagen: Author.

Fine, B., & Leopold, E. (1993). *The world of consumption.* London: Routledge.

Galli, S., Scaglione, A., & Wang, Z. (2011). For the grid and through the grid: The role of power line communications in the smart grid. *Proceedings of the IEEE, 99*(6), 998–1027. doi:10.1109/JPROC.2011.2109670

Gontar, B., Gontar, Z., & Pamuła, A. (2013). Deployment of Smart City Concept in Poland. Selected Aspects *SIsteminiai Tyrimai, 67*(3), 39-51. 10.7220/ MOSR.1392.1142.2013.67.3

Hamilton, B., Miller, J., & Renz, B. (2010). *Understanding the benefits of smart grid.* Tech. Rep. DOE/NETL-2010/1413. U.S. Department of Energy.

Harvey, D. (2005). *Brief History of Neoliberalism.* New York: Oxford University Press.

Hughes, T. (1983). *Networks of power: Electrification in Western society.* Baltimore, MD: Johns Hopkins University Press.

IEA. (2009b). *Comparative Study on Rural Electrification Policies in Emerging Countries.* Paris.Available at http://www.iea.org/papers/2010/rural_elect.Accessed 18/7/2017

Johnson, P. (2009). *Human centered information integration for the smart grid*. Technical Report CSDL-09-15, University of Hawaii, Honolulu, HI.

Kannberg, L., Kintner-Meyer, C., Chassin, D., Pratt, R., DeSteese, J., Schienbein, L., . . . Warwick, W. (2003). *Grid Wise: The Benefits of a Transformed Energy System*. Pacific Northwest National Laboratory under contract with the United States Department of Energy: 25. arXiv:nlin/0409035.

Lehmann, S. (2010). Green Urbanism. *Formulating a Series of Holistic Principles Surveys and Perspectives Integrating Environment and Society*, *3*(2), 1–10.

Lichtenberg, S. (2010). *Smart Grid Data: Must There Be Conflict Between Energy Management and Consumer Privacy?* National Regulatory Research Institute.

Mazza, P. (2004). The Smart Energy Network: Electricity's Third Great Revolution. *Climate Solutions*, *1*(2), 1–7.

Milner, D. (2012). *Consumption and Its Consequences*. Cambridge Polity Press.

Mulgan, G. (1991). *Communication and Control*. Cambridge, UK: Polity Press.

Pacific Northwest National Laboratory. (2007). *Gridwise History: How did GridWise start?* Available at http:/gridwise.pnl.gov/foundations/history.stm Accessed 20/9/2017

Register, R. (1987). *Ecocity Berkeley: Building Cities for a Healthy Future*. North Atlantic Books.

Rennings, K., Brohmann, B., Nentwich, J., Schleich, J., Traber, T., & Wüstenhagen, R. (Eds.). (2013). *Sustainable Energy Consumption in Residential Buildings*. Springer. doi:10.1007/978-3-7908-2849-8

Ronfeldt, D. (1992). Cyberocracy is coming. *The Information Society*, *8*(4), 243–296. doi:10.1080/01972243.1992.9960123

Saleh, M., Althaibani, A., Esa, Y., Mhandi, Y., & Mohamed, A. (2015). Impact of clustering microgrids on their stability and resilience during blackouts. *2015 International Conference on Smart Grid and Clean Energy Technologies (ICSGCE)*, 195–200. 10.1109/ICSGCE.2015.7454295

Sassen,S .(2005). The Global City: introducing a Concept. *The Brown Journal of World Affairs*, *11*(2), 27-43.

Schor, J. (2001). *Why Do We Consume So Much?* Clemens Lecture Series 13 Saint John's University.

Schor, J. (2005). Prices and quantities: Unsustainable consumption and the global economy. *Ecological Economics*, *55*(3), 309–320. doi:10.1016/j.ecolecon.2005.07.030

Shove, E. (2003). Converging conventions of comfort, cleanliness and convenience. *Journal of Consumer Policy*, *26*(4), 395–418. doi:10.1023/A:1026362829781

Spaargaren, G. (1997). *The ecological modernisation of production and consumption: in environmental. sociology*. Wageningen, The Netherlands: Landbouw University Wageningen.

Speake, S. (2005). Water:A Human Right. In *Consuming Sustainability Critical Social Analysis of Ecological Change*. Fernwood Publishing.

Succar, S., & Cavanagh, R. (2012). *The Promise of the Smart Grid: Goals, Policies, and Measurement Must Support Sustainability Benefits*. NRDC Issue Brief.

Thompson, C. (1996). Caring consumers: Gendered consumption meanings and the juggling lifestyle. *The Journal of Consumer Research*, *22*(4), 388–407. doi:10.1086/209457

UNEP/Wuppertal Institute Collaborating Centre on Sustainable Consumption and Production (CSCP). (2005) *Sustainable Energy Consumption*. A Background Paper prepared for the European Conference under the Marrakech Process on Sustainable Consumption and Production (SCP), Berlin, Germany.

U.S. Department of Energy. (2003). *"Grid 2030" A National Vision for Electricity's Second 100 Years*. Office of Electric Transmission and Distribution.

U.S. Department of Energy. (2004). *National Electric Delivery Technologies Roadmap*. Office of Electric Transmission and Distribution.

US Energy Independence and Security Act 2007 (EISA-2007)

Warde, A. (1999). Convenient food: Space and timing. *British Food Journal*, *101*(7), 518–527. doi:10.1108/00070709910279018

WCED (World Commission on Environment and Development). (1987). Our Common Future. Oxford, UK: WCED.

Werner-Allen, G., Lorincz, K., & Marcillo, M. (2006). Deploying a Wireless Sensor Network on an Active Volcano. *IEEE Internet Computing*, 1-25.

World Economic Forum. (2016). *Digital Transformation: Digital trends in the automotive industry*. Available at http://reports.weforum.org/digital-transformation/digital-trends-in the automotive-industry.Accessed 9/21/2017

Yang, Q. (2001). *Electrical engineering and its automation*. Beijing Sifang Automation Co. Ltd.

Young, D. (2008). When do energy-efficient appliances generate energy savings? Some evidence from Canada. *Energy Policy*, *36*(1), 34–46. doi:10.1016/j.enpol.2007.09.011

Yu, F., Zhang, F., Xiao, W., & Choudhury, P. (2011). Communication Systems for Grid Integration of Renewable Energy Resources. *IEEE Network*, *25*(5), 22–29. doi:10.1109/MNET.2011.6033032

Zhao, J., Huang, W., Fang, Z., Chen, F., Li, K., & Deng, Y. (Eds.). (2007). *Proceedings of Power Engineering Society General Meeting*. doi:10.1109/PES.2007.385975

Zuboff, S. (1988). *In the Age of smart machines*. New York: Basic Books.

KEY TERMS AND DEFINITIONS

EISA: Energy Independence and Security Act.

ISDN: Integrated services digital network.

ISO: International Organization for Standardization.

LSI and VLSI: Large scale integration and very large scale integration. The integration of large numbers of computers on a single microchip.

MAP: Manufacturing automation protocols.

ONA: Open network architecture.

ONP: Open network provision, set of proposals devised in 1987 by European Commission to allow for open access to the communication infrastructure.

SNA: Systems network architecture.

Chapter 6

Future Directions to the Application of Distributed Fog Computing in Smart Grid Systems

Arash Anzalchi
Florida International University, USA

Longfei Wei
Florida International University, USA

Aditya Sundararajan
Florida International University, USA

Amir Moghadasi
Florida International University, USA

Arif Sarwat
Florida International University, USA

ABSTRACT

The rapid growth of new technologies in power systems requires real-time monitoring and control of bidirectional data communication and electric power flow. Cloud computing has centralized architecture and is not scalable towards the emerging internet of things (IoT) landscape of the grid. Further, under large-scale integration of renewables, this framework could be bogged down by congestion, latency, and subsequently poor quality of service (QoS). This calls for a distributed architecture called fog computing, which imbibes both clouds as well as the end-devices to collect, process, and act upon the data locally at the edge for low latency applications prior to forwarding them to the cloud for more complex operations. Fog computing offers high performance and interoperability, better scalability and visibility, and greater availability in comparison to a grid relying only on the cloud. In this chapter, a prospective research roadmap, future challenges, and opportunities to apply fog computing on smart grid systems is presented.

DOI: 10.4018/978-1-5225-3996-4.ch006

Copyright © 2018, IGI Global. Copying or distributing in print or electronic forms without written permission of IGI Global is prohibited.

INTRODUCTION

Unlike conventional power systems, smart grid today is a healthy amalgamation of multiple interoperable, scalable, efficient, sustainable and secure technological domains contributing holistically towards the availability, reliability, and quality of the power generated, transmitted, and distributed to consumers. These technological domains extend beyond power systems themselves, manifesting as data management, cyber-physical security, human behaviors, mathematics, communication, and even wireless sensors. Such an interconnected smart grid can provide sustainable and reliable power delivery, ensuring availability and eco-friendly means of generating power. Integrating many renewables into the grid, at both transmission as well as distribution levels, is expected to yield an unstable grid owing to their intermittent nature. This necessitates the need for smoothing as well as optimizing the cost of operation, production, and distribution (Popeanga, 2012). Consumers have now metamorphosed into "prosumers", wherein they have gained the ability to both produce as well as consume power. With the integration of more intelligent sensors and devices on the field, largely distributed across the grid, Internet of Things (IoT) has gained prominence in this critical infrastructure as well. These devices churn data points constantly, racing to the central servers bolstered by strong and resilient communication infrastructure (Dastjerdi & Buyya, 2017). Consequently, the significance of computing in the smart grid domain must be investigated. For this purpose, three sub-domains within the smart grid will be explored in this chapter: information management, energy management, and security.

Significance of Computing in Information Management

Currently, the deluge of digital data is centrally ingested and cleansed into useful information blocks that can be processed. Many critical power system applications such as Demand Response (DR), load flow and Optimal Power Flow (OPF) analyses, fault and reliability analyses, quality assurance models, customer billing processes, direct load control, active/reactive power control and distribution planning, to name a few, depend on the veracity and validity of the information gathered from the field. This warrants the need for effective and powerful computing technologies for smart grid.

Significance of Computing in Energy Management

The existing power grid needs an optimal balance of electricity demand and supply between consumers and the utilities. The smart grid can address this requirement. Such features in a smart grid are realized by the integration of various Energy

Management Systems (EMSs) such as Home Energy Management (HEM), Demand Side Management (DSM), and Building Energy Management Systems (BEMS) (Fang, Misra, Xue, & Yang, 2012). A smart grid allows various renewable energy sources (such as solar and the wind) to have the efficient management of supply and demand. In the emerging smart city and microgrid scenarios, EMS is one of the most important cornerstones. Naturally, for the efficient and continuous operation of such complex ecosystems within the grid, computation will play a crucial role. With the integration of secondary storage devices such as batteries and supercapacitors, optimal economic dispatch, peak load shaving, and other operations will be required at micro-grid level. Under such scenarios, effective distributed computing methods need to be added on top of the underlying power infrastructure.

Significance of Computing in Security

Smart grid today is a Cyber-Physical System (CPS). While historically air-gapped with cyber and physical silos, the modern grid comprises multiple interdependent elements from both cyber as well as physical realms (Harp & Gregory-Brown, 2013). The cyber realm comprises the Information Technology (IT) solutions that include data processing servers, networked operator consoles, and workstations, visualization interfaces as well as the communication infrastructure itself. On the other hand, the physical realm contains the Operation Technology (OT) solutions that include Industrial Control Systems (ICSs) like Supervisory Control and Data Acquisition (SCADA), Process Control Domains (PCDs), Distributed Control Systems (DCSs) and Programmable Logic Controllers (PLCs). While IT relies heavily on networking functionalities to pull data from different sources and visualize useful insights for the operators and analysts continuously, the OT adopts proprietary protocols and the ICSs for control and dispatch. This silo-based operating framework, however, is no longer adequate, considering the proliferation of more intelligent field sensors that discourage the concept of such a framework and are required to interweave the principles of IT and OT. However, this increased intermingling between IT and OT realms has exposed the vulnerabilities of the grid's critical infrastructure elements greater than ever before. It cannot be denied that a cyber-attack on the modern grid will have definite ramifications in the physical realm, and vice-versa. Recent successful cyber-attacks or even cascading failures on power grids across the world, such as the 2003 Northeastern United States blackout due to cascading failures (Chadwick, 2013; National Electric Reliability Commission [NERC], 2004), the 2015 Ukrainian grid blackout due to BlackEnergy3 malware (Electricity Information Sharing and Analysis Center 2016), and attacks on SCADA/ICS due to the HAVEX worm (Nelson, 2016) bear testimonies to this fact.

Most attacks target sensitive IT gateways and checkpoints, including operator workstation consoles and emails, through methods such as phishing, malware, social engineering or Denial of Service (DoS) by exploiting vulnerabilities like poor security configuration, lack of situational awareness, susceptibility of ill-informed or poorly-aware system operators, bugs in critical OT-managing software, poorly maintained Access Control Lists (ACLs), and inadequate patching and firmware updates (Wang, 2013). In response to these concerns, the National Electric Reliability Commission (NERC) has put forth several Critical Infrastructure Protection (CIP) standards for cyber-security, including those for personnel and training, incident reporting and response planning, information protection, physical security and vulnerability assessment (NERC, 2012, 2014a, 2014b, 2014c). It is imperative that to successfully ensure adherence to these standards, effective computational technologies and methods must be in-place.

Contribution of This Chapter

While existing computation tools increasingly imbibe the principles of Cloud Computing (referred to as *Cloud* moving forward), its centralized processing is counterproductive to the inherent way the smart grid is designed to function, which is distributed. The energy infrastructure requires high performance, lightweight, low latency, and distributed intelligent methods which shift significant processing and management workloads from the center towards the edge of the grid. The emerging applications that involve computations directly or indirectly need to inculcate an optimal combination of computing principles, giving way to the new paradigm of "Fog Computing", referred to in this chapter as *Fog*. Although Fog is touted to be equivalent to Edge Computing (referred to moving forward as "Edge"), there exists a little difference. While the Cloud looks at retaining its intelligence only at the central point of an infrastructure, the Edge attempts to move significant computations to the infrastructure's end devices. However, the Fog tries to strike a balance by shifting low latency, lightweight applications to the edge and high latency, heavyweight applications to the Cloud. In this case, Fog can be thought of as a combination of the Edge and the Cloud, and under such a context, the term "Fog" becomes interchangeable with "Edge". In this chapter, the word "Edge" will be used in this context unless otherwise stated.

Fog addresses the inherent drawbacks of the Cloud such as latency, energy consumption concentration, bandwidth and associated costs, and privacy. It also presents additional benefits such as localization, location-awareness, operational visibility and dynamic resource management, and fulfills business goals (Yi, Li & Li, 2015). In addition, it is predicted that the Fog utilization will have a strategic role in the motivation and design of the emerging smart grid and will increase the robustness of its communication. This chapter, hence, lays the much-needed emphasis on the distributed fog computing architecture and framework, which was initially coined by Cisco and later structured and disseminated by the OpenFog Consortium in 2015, of which Cisco is also a member (Cisco, 2015). The authors believe that this chapter will serve as a unique starting point for readers interested in further investigating the potential of Fog Computing on the future smart grid for better performance, stability, and reliability.

Organization of This Chapter

In Section II, a foundation is first laid by reviewing the traditionally employed computing tools and contrasting them with the more recent cloud computing technologies for the smart grid. Applications of Cloud to energy management, information management, and security are discussed. Section III introduces the envisioned Fog architecture for the smart grid, which involves a two-tiered approach comprising Local and Central Tiers. The introduced framework's significance and validity in energy management, information management, and security is then investigated by the authors. Section IV presents a case study to provide the context for mechanisms included in Fog by considering different power system applications that can be then validated, such as short-term Distributed Renewable Resource (DRR) generation forecasting, unit commitment and adequacy assessment for scenarios involving high penetration DRRs. The mathematical models for these applications are provided first, followed by an explanation of how the envisioned Fog model can be used to achieve their implementation. Following that, in Section V, the proposed testbed simulator is introduced and described. Further, it is shown how this testbed is used for validating the applications discussed in Section IV.

BACKGROUND: OVERVIEW OF SMART GRID AND EXISTING COMPUTING

Before introducing and discussing the proposed Fog framework, it is important to first discuss the smart grid architecture that lays the foundation for the computing models discussed in this chapter. Then, the emergent Cloud framework must be briefly discussed, since it has the precursors to Fog. To help the reader better appreciate the pros and cons of these two technologies, they are explained in the context of three major power system applications: energy management, information management, and cyber-physical security, which are also covered in the smart grid architecture discussion.

Smart Grid Architecture

It is, by now, clear that smart grid cumulatively generates continuous streams of high-dimensional, multi-variate data-points which bear the characteristics of big data: large volume, high velocity, acute veracity, diverse variety and critical validity, collectively called the "Five Vs" of smart grid big data. Although these characteristics are applicable to all forms of big data, they are more crucial for smart grid because of their direct implications on power delivery and security, as briefly investigated in Section I. Data volume refers to the size of generated data which is generally of the orders of Terabytes (TB) or Petabytes (PB) every month in most utility-operated territories. Data velocity attributes the speed with which these devices churn data observations, which ranges from the orders of a few milliseconds to one hour. Velocity can also be understood as a measure of data resolution. This augments with the concept of variety, which signifies the different sources that spawn data records. Typical sources of today's automated data generation include the Advanced Metering Infrastructure (AMI) smart meters, Phasor Measurement Units (PMUs), meteorological weather stations, and Intelligent Electronic Devices (IEDs) such as Voltage Regulators, Load Tap Changers (LTCs) and Capacitor Banks. Data veracity encapsulates the idea of data integrity and consistency, which is important for the smart grid, considering important decisions regarding power dispatch and demand response are made based on processes which use these data. Finally, data validity represents the freshness of data (Sornalakshmi & Vadivu, 2015). There are scenarios in power systems where computational algorithms must process a newly generated data almost in real-time to generate actionable results. A delay even by one minute could cause significant losses in revenue or energy. Such latency-cognizant applications need to have constant and on-demand access to data.

A conceptual architecture of the future grid in smart cities is shown in Figure 1. It comprises the physical devices and controllers at the edge level such as AMI meters, PMUs and IEDSs among others. These devices and electronic systems communicate with the Edge Computing module of the Fog, forming the Local Tier. Simple analytics and security protocols are executed at this level, with results aggregated and encapsulated with partially processed data to increase abstraction. Using designated gateways, over an efficient and available network connectivity, further bolstered by optimal routing strategies, the information is forwarded higher in the hierarchy on a need-to-compute basis. Diagnostic analysis methods such as contextualization, association mining, correlation, simple optimization and iterative processes are executed at this point. It is key to remember that as the data goes higher in the hierarchy, the processing power and scale grows rapidly as well. Thus, the nature of the data itself morphs into one that of big data. As it enters the Central Tier where the Cloud still resides, heavyweight applications come into picture. Most of

Figure 1. The smart grid architecture and its tiers of computing

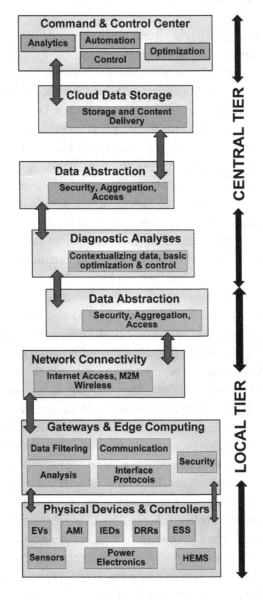

these applications are legacy, being run at the utility's CCC. The applications at this level must have immense computational power, ability to run parallel processes over multiple cores, and fully exploit the robustness and flexibility of the cloud platform. It is noticeable that data gets proliferated across the two Tiers in multiple ways, thus increasing ubiquity and connectivity. This in-turn increases the attack surface of the overall smart grid. Hence, cybersecurity approaches also need to be decentralized, implemented at varied complexities in the Local as well as Central Tiers.

Cloud Computing Framework

The cloud infrastructure can empower customers to have immediate access to the data they are entitled to, via customized applications, from anywhere anytime, provided they are connected to the network. Recently, the use of cloud computing for managing and processing smart grid big data has been discussed in the literature. Nozaki, Tominaga, Iwasaki, and Takeuchi (2011) suggest administering central optimization frame and communication infrastructure using cloud computing data hubs as auxiliary cognitive radio system of smart meters. Fang, Misra, Xue, and Yang (2012) proposed a model of data management for the smart grid based on cloud computing is offered. This model exploits distributed administration for instantaneous data acquisition, recovery of real-time information as a parallel analyzer, and network-wide access. A system for decision support as well as a software for cloud computing is offered by Guo, Pan, and Fang (2012). This technology brings together consumers, energy consultants, modern web interoperable technologies, and energy service procedures. Some studies compare smart grid with the traditional power grid to recognize new weaknesses and exposures in the grid of the future. Zaballos, Vallejo, and Selga (2011) proposed an architecture for private cloud computing which can be used for supporting smart grid. An agent-based cloud-client architecture which is designed for the smart grid is introduced by in Metke and Ekl, (2010). This technology can combine storage resources and computing, and make no changes to the interior configuration or current equipment of the system. A systematic approach for smart grid virtualization at both device and substation levels is introduced. A high-level overview of devices, trust, and communication security, as well as complexity and scalability issues of the smart grid, is presented by Zhang, Wang, Sun, Green, and Alam (2011).

Energy Management

Several cloud computing methods have been proposed in the literature for energy management in the smart grid. Today, most of the utilities have an EMS which emphasizes instantaneous grid management. Global smart grid framework needs accessible software infrastructure which analyses and integrates flowing of real-time data from numerous smart meters for optimization and investigation of supply and demand curves. Cloud platforms are being well studied to reinforce such enormous data and compute-intensive, online applications. For these purposes, scalable constraints are obtainable by the cloud products to shape a dynamic and real-time software. Cloud as an essential part of this environment offers these benefits:

- The elastic operation of cloud helps to avoid inflated capital cost at the peak load, therefore it helps to cut the expenses of the utilities.
- Real-time pricing statistics and power consumption data can be shared with customers and they can take advantage of the real-time information.
- Information can be securely distributed to third party organizations through cloud facilities, in order to build smart tools to tailor end user needs.

Information Management and Communication

Currently, numerous advanced smart meters are implemented at the distribution grid. Cloud computing is a beneficial service to effectively store and process this bulky data for information management, for these reasons:

- Smart grid data analysis process matches the storage and handling mechanism of the cloud.
- In a smart grid, data distribution is a significant concern in the smart grid structure. Various can be pooled by implementing cloud computing.
- Main players of the smart grid, e.g. consumers, utilities, and microgrids, can get access to the cloud data, even though they are operating in islanded mode.
- A significant characteristic of the smart grid is two-way information and energy flow which empowers several functions and operational control systems in the smart grid, the however real-time process of smart grid behaviors can be done in the cloud platform.
- Management of massive data is complex, costly, and may be beyond the capacity of existing data management systems in the smart grid. Using a cloud-based information management system can help in encountering such drawbacks.
- A cloud-based information center can help managing huge data in a cheaper way. Moreover, the current system's ability to do such a thing have much more limitations than cloud.

Cyber-Physical Security

Smart grids can be assumed as a cyber-physical scheme which makes the physical systems (electricity) function in parallel with cyber-framework (internet). This facility as the backbone of the can communicate with the consumer appliances and also provide the backbone for service providers to absorb contents and control operations. With the presence of online connectivity, it is a big challenge to prevent cyber-attacks in the smart grid that can potentially disrupt the power supply.

To solve these problems in the smart grid expansion, a number of security mechanisms have been proposed in the literature by application of cloud computing, which are listed below:

- A security structure for power system and protection system information, established on the security of the cloud, is offered by (Yanliang, Song, Wei-Min, Tao, and Yong (2010). Security of the cloud was separated into two sections: client and server. Data are collected by the client and the actions are chosen by the server's reactions. On the other hand, On the other hand, the cloud computing is exploited by the server as a smart decision maker and distributed storage. After that, by use of the Internet, the results are transmitted.
- The security of stored data in cloud computing is guaranteed by the distributed verification protocol (DVP). The execution of the DVP protocol is appropriate for power management and data storage structures in the smart grid (Ugale, Soni, Pema, & Patil, 2011).
- Many security threats are presented by Yang, Wu, and Hu (2011), such as (a) the personal sensitive data can be accessed by the cloud supervisors, (b) security of transmission and storage of the data is in a close relationship with the cloud location.

THE FOG COMPUTING FRAMEWORK

Smart grid big data can be grouped into four broad categories: Operational, Non-operational, Event message, and Meter data. Operational and non-operational datasets are part of the grid's Energy Management System (EMS) and Distribution Management System (DMS). While Event message data corresponds to Outage Management System (OMS), the meter data belong to the Meter Data Management (MDM). These logical zones of the grid are meant to be interoperable and constantly exchange data between their layers for executing different applications. The operational data are concerned with voltage and current phasors, real and reactive power flows, DR, OPF constraints and values, and generation forecasts. Non-operational data corresponds to data on asset health, power quality, reliability, asset stressors, and telemetry. Event message data is mostly concerned with asset vulnerabilities, threats, failures, and likelihood of risks including but not limited to device flags, checksum errors, fault detection events, device logs, alert messages, and meter voltage loss information. Meter data includes the energy consumption readings, the average peak and time of day values.

The Fog framework supports an architecture that enables the processing of latency-sensitive data and computations closer to the sources of such data, instead of sending them unprocessed all the way to the cloud and retrieve results. As shown in Figure 2, the Fog framework combines both the Edge and the Cloud in a manner that increases latency top-down. At the edge of the grid are the various data source nodes which constantly generate data-points which fall into one of the four categories described in Section II. These nodes, depending on the volume and validity of the data they generate, are allocated dedicated or shared computation-rich processing nodes called "Fog nodes", all of which, along with the source nodes themselves, form the *Local Tier*. The Fog nodes could be cloudlets, smartphones, Personal Digital Assistants (PDAs) or even client computers. Latency-intensive, stream computations including preliminary descriptive analytics on the raw data could be executed at this Tier (Sajjad, Danniswara, Al-Shishtawy & Vlassov, 2016). For example, in the AMI architecture, a few hundred residential smart meters feed their data continuously to an Access Point (AP). A Fog node dedicated to this AP could check the data's quality and integrity, clean erroneous values, identify missing records and correlate them with the checksum values and device flags to create a rudimentary analysis. Data obfuscation and privacy-preserving policies could also be implemented at this Tier before the node forwards the semi-processed data to the node next in the hierarchy. Depending on the application, there could be more than one level of Fog nodes in the grid, where the data is subject to processing at each level until it reaches the Cloud, called the *Central Tier*. This Tier might not only be the central cloud. For example, an industry might have its own private cloud with which its Fog nodes interact, and the private cloud finally sends only the required information to the publicly managed Cloud at the center, where very heavyweight computations employing batch processing, process optimizations, state estimation and network topology analysis could be performed on the aggregated data.

As can be seen above, Fog framework has a dynamic topology, where different IoT devices could interact in different ways depending on their needs and business model requirements. This, unlike in a Cloud environment, raises significant concerns, elaborated in the following subsection.

A. How Fog Differs From the Cloud?

It is now time to examine the key areas where the Fog and the Cloud exhibit differences. The observations, gathered mostly from industrial white papers and technical articles in journals and proceedings, are summarized and tabulated below in Table 1 (Rao, Khan, Maschendra & Kumar, 2015; Saharan & Kumar, 2015; Yi, Hao, Qin & Li, 2015; Almadhor, 2016; Deshmukh & More, 2016; Okay & Ozdemir, 2016; Luntovskyy & Spillner, 2017). It can be concluded from this comparison that

Figure 2. The envisioned Fog Computing Framework for Smart Grid

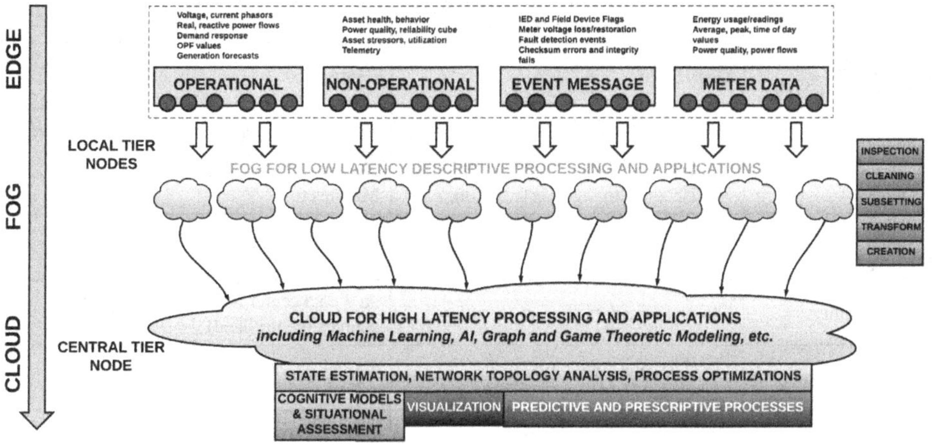

there are meritorious aspects to the Fog that the Cloud cannot compete against. At the same time, there are aspects where the Cloud has a winning hand over the Fog (Shi & Dustdar, 2016). Hence, application of Fog to one's requirement is not universally justifiable and is greatly dependent on one's business and operational requirements. However, Fog has generally found a greater applicability than the Cloud for any IoT or IoT-like ecosystem, of which smart grid is increasingly becoming a part.

The Architecture

Fog computing is introduced and defined by the OpenFog Consortium (2017) as "a horizontal, system-level architecture that distributes computing, storage, control and networking functions closer to the users along a cloud-to-thing continuum" (p. 3). This definition is holistic considering it advocates the distributed nature of Fog and the fact that it integrates the individual concepts of the Cloud and Edge, applicable the best for IoT environments. Another noteworthy aspect is its emphasis on just computing being distributed, but also storage, control and networking (communication). This implies that the Fog is a heterogeneous ecosystem where constituent devices need to be interoperable and cross-functional. It is quite different to the notion of a Cloud, where all devices engage in a unilateral correspondence with the central nodes and rely on them for all their needs. Keeping in mind these key views, the Consortium proposed the OpenFog Reference Architecture (OpenFog RA) to help various Fog stakeholders like managers, developers, and users (OpenFog Consortium, 2016). For consistency, the OpenFog RA is referred to in this chapter as the Fog architecture.

Table 1. A high-level comparison between the Fog and Cloud frameworks

	The Fog	The Cloud
Computing Architecture Style	Distributed	Centralized
Suitable for	Lightweight, moderate-weight applications requiring low latency	Heavyweight applications that can survive medium to high latency
Latency	Milliseconds to seconds (real-time)	Minutes to days
Information Storage	Streaming (hours to few days)	Batch (months to years)
Data locality	Closer to the data sources (localized)	At a data center located away from the sources (centralized)
Framework Scope	Local to each source (device/ aggregator-level)	Global scope (network or infrastructure-level)
Heterogeneity	High (different sources need to be inter-operable across varied protocols, schemas, and structure)	Low (nature of original sources does not matter as data is pooled and centrally managed)
Interactions	Machine-to-Machine (M2M)	Machine-to-Command and Control-to-Machine (MCM)
Common Application	Descriptive analytics, local control	Predictive, prescriptive analytics, global control
Partitioning	Information and applications are partitioned between Edge and Cloud for optimal processing	Information and applications are partitioned within the Cloud for optimal storage, security
Privacy	By virtue of data localization, privacy can be better guaranteed	Under public Cloud environments, privacy is still a debated concern
Maintenance	Owing to heterogeneity in protocol and topology, maintenance is difficult	Owing to its central storage and processing, maintenance is easier

Now that the envisioned distributed fog framework has been introduced, it is time to investigate the different characteristics of the framework, namely the factors that define the Fog architecture. They also help in evaluating the performance and applicability of this architecture to a critical infrastructure like smart grid. The Fog architecture can be described by nine key characteristics which are inter-dependent and complementary in nature (Byers & Wetterwald, 2015). They have been briefly summarized below, but the authors regard them as important emerging and open areas of research.

1. **Reliability:** Like other traditional computing paradigms such as mainframe, grid, cluster and cloud, the Fog also must satisfy some of the critical standard reliability characteristics with respect to physical and computational resources (Madsen, Albeanu, Burtschy & Poentiu-Vladicescu, 2013). While physical

resources include computing hardware, sensor nodes, actuator nodes and augmented smart technologies like smart inverters, IEDs and PMUs to name a few, computational resources include the models and algorithms running over the software platforms. The Fog derives characteristics from its predecessors, such as dynamic resource allocation and deallocation (see C.4), parallel processing (through its distributed nature) and virtualization. The authors refer to Madsen et al. (2013), which considers the failure of sensor nodes, lack of network coverage, failure of the platform or the network itself, or the failure of just the user interface as part of Fog reliability.

2. **Hierarchical Structure:** By definition, the Fog nodes of a particular Local Tier are designed to be as close to the end-devices as possible. For a simple infrastructure, there could be one Local Tier and one Central Tier. However, for more complex networks like the smart grid, there are bound to be multiple Local Tiers that retain operational autonomy but are interoperable with their peer Tiers. The Local Tiers higher up by a step could aggregate two or more end Local Tiers which are closely related in terms of functionality or resources. For example, in a smart community, different Local Tiers corresponding to micro-grids, the utility's distribution network, AMI meters and autonomous Electric Vehicles (EVs) can exist. These Local Tiers almost always need to interact with one another since their unimpeded operations depend on applications such as energy management, power flow analysis and direct load control. For this purpose, a medium-level Local Tier could be built which aggregates relevant information from the end-Local Tiers and handles medium-weight processes. The Central Tier (Cloud) could then conduct demand response, load profiling and customer billing processes based on the results delivered by the different layers of Local Tiers. This hierarchical, multi-tier structure of the Fog makes it quite adaptive in nature, which cannot be achieved by Cloud.

3. **Resilience:** Resilience is an attribute of how effectively the mission-critical applications continue to run even in the event of a failure or an attack. Being distributed not only in computing, but also in storage, processing and communication, Fog architecture is one of the best approaches to ensure resiliency (Pradhan, et al. 2016). The Fog nodes are meant to be dynamic, interoperable, and hierarchical in nature. Should a Local Tier be compromised, resources for the nodes in that Tier could duly be deallocated and reallocated to the nodes in adjoining Tiers which can take over the additional load, which in-turn can be smartly distributed to multiple peer Tiers based on their own computation loads. Should the Central Tier itself be compromised, all the Local Tiers can still run critical applications as usual since they do not depend on the Cloud for them. Fog nodes usually employ checkpoints which identify the most recent "safe" mode of operation, to which they could rollback upon

restoration. Replication or redundancy is another tolerance mechanism which is more applicable for a distributed framework like Fog, considering the enormity of information to be replicated at a Local Tier is much less than at the Central Tier in the event of a disaster or attack.

4. **Dynamism and Heterogeneity:** The hierarchical structure of a typical Fog framework was discussed. It follows from this characteristic that the Fog is also dynamic in nature. Due to the varying energy needs dictated by dynamic load profiles, for instance, the intermittent generation profiles of renewables installed in the micro-grid, and the inclusion of mobile smart loads such as EVs, there could be scenarios that warrant idle processing power, skewed loading of Fog nodes both within a Tier and across Tiers, etc. Sometimes, physical and computational resources become surplus or deficit. The Fog architecture is capable of dynamic allocation, deallocation and reallocation of resources to nodes across layers (Emfinger, Dubey, Volgyesi, Sallai & Karsai, 2016). The added heterogeneity of the original sources implies greater dynamism.

5. **Partitioning:** The Fog also supports the principle of data and application partitioning, where certain parts of the data generated by the source nodes are ingested at the Edge while the remaining are reserved for the Cloud. A similar concept is applicable to the applications that depend on these data. Consequently, they require to be partitioned in a manner that Fog-data is utilized by Fog-Applications and Cloud-data by the Cloud-applications (Khedkar & Gawande, 2014).

6. **Interoperability, Automation and Autonomy:** Fog is meant to be transparent and support ubiquitous computing. Interoperability of the Fog resources is critical to ensure seamless Machine-to-Machine (M2M) communications at the Tiers. The architecture also enables autonomy of resources and services for discovery and registration, management, allocation and deallocation, security (such as Authentication-Authorization-Accounting or InfoSec), and most importantly, operation. Autonomy and self-healing nature are considered the defining traits of the smart grid, and the Fog is strategically positioned to facilitate moving a step closer to fully realizing this goal. The architecture further supports the use of Application Programming Interfaces (APIs), standards and containerization (isolated execution for looser coupling) of applications for customizing nodes and Tiers, through which operational dynamism and interoperability can be automated. Automation can be additionally achieved by implementing Software-Defined Network (SDN) technologies and M2M communications like Message Queuing Telemetry Transport (MQTT) and Constrained Application Protocol (CoAP) also lend towards increased automation and interoperability. In some cases, SDN-managed Network Functions Virtualization (NFV) can help reduce

physical resource constraints and enhance load management (Gupta, Nath, Chakraborty & Ghosh, 2016).

7. **Latency and Bandwidth:** Probably the most talked about characteristic of the Fog is its low latency and high utilization of available bandwidth. It is expected that the Fog nodes within Local Tiers are energy-conservative in nature. Hence, they are resource-constrained and bandwidth-aware (Borcoci, 2016). Since many Fog nodes handle time-critical and mission-critical analyses, factors such as data validity is important. For events such as fluctuating renewable energy generation due to frequently passing clouds that might negatively impact the grid it is connected to, the associated Fog nodes have to not only smoothen the spikes and dips but must also be capable of accurately predicting these intermittencies so that secondary generation dispatch can be scheduled. For this, data validity is key, as basing results on outdated data could lead to erroneous forecast since the renewable generation values are bound to change rapidly under high penetration levels envisioned in the future. This consequently boils down to computational latency, communication latency and network bandwidth. While the first factor is dependent on the node's memory, processing power and the application itself, the second and third factors depend on the topology of the Fog, availability of resources and nature of data.

8. **Security and Privacy:** Privacy, although technically different from security, can be considered part of the whole. Fog architecture is founded upon an end-to-end approach where the nodes of the grid can interact with one another at a local level and only avail the services of a cloud when a problem requires the solution from a heavyweight computation module. Even when Fog nodes are deployed as Fog as a Service (FaaS), security is built through the concepts of Root of Trust (RoT), authorization and privacy. It is expected of Fog nodes to establish trust with other nodes and underlying assets they are tied to, by discovering and maintaining a chain of authorization, termed formally as attestation. The node closest to the end-device within the Local Tier could be the first point of security. Before passing this protected data to the next level, the node must attest its trust. As the continually processed data is handed higher up both within the Local Tiers and to the Central Tier, the nodes each add their attestations, thus establishing a chain of authorization. Owing to the dynamism of the nodes in the Fog, it is required that the physical and computational resources also be attested along with data.

9. **Agility:** It is, by now, well understood that data by itself has no meaning. For it to obtain meaning, it needs to be duly ingested, processed, computed and turned into useful information. The missing link in bringing this full circle is "context", which the Fog is also capable of achieving. Prior to processing, the

original data is spurious, poorly structured, haphazardly encoded and might even contain a few missing values. This data as such is unfit for analysis and requires undergoing a series of processes, collectively called descriptive analytics, as was discussed in the introduction of Section III. However, as important as it is to format and cleanse the data, it is crucial to derive appropriate context for it as well, since most computation processes depend on data's context for providing results. The contextual understanding must expedite the ability of the models to make faster and optimal decisions that then drive broader decisions and policy management higher in the hierarchy. Once a context is established, the applications can be executed on any Fog node located on any layer of the architecture, and hence become more robust to every-day changes in data patterns or system behaviors.

APPLICATION OF THE FOG TO DEMAND MODELING FOR NETWORK

An overview of the proposed fog framework is shown in Figure 3, which considers a specific use-case of demand modeling to explain its significance. Power supply and demand in a grid are continuously changing. The challenge has become far fiercer with higher penetration of intermittent renewables. Demand modeling has taken a new turn with novel technologies to collect almost real-time load information.

Figure 3. Schematic view for demand modeling at central and local tiers of the proposed fog framework

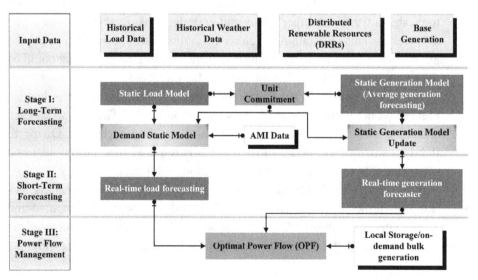

Demand modeling involves understanding and formulating complex mathematical models for customer behavior. The electricity demand curves are ever changing with the introduction of newer kinds of loads. To obtain a comprehensive load modeling, a two-tier fog framework is introduced. In the future smart grid, demand modeling would be required at both Local as well as Central Tiers. For example, demand modeling for a micro-grid in a community can be considered at the Local Tier, while the modeling for a larger community (like a city) falls under the Central Tier. In both scenarios, a common set of steps need to be executed. The first step utilizes historical load and weather data to achieve a platform for future dynamic load identification. The second step upgrades this model using data from Advanced Metering Infrastructure (AMI) and Intelligent Electronic Devices (IEDs). As a corollary, historical and 15 minute-interval data will be utilized to conduct large-scale studies to develop a model to predict the behavior of individual customers and then aggregate the predicted models to achieve a comprehensive demand pattern.

A similar approach is adopted on the generation side, where the available power from renewables can be accurately estimated only for shorter duration of time. To model the uncertainty of renewable resources, two main factors should be considered: 1) weather prediction, 2) the performance of the solar panels and wind generators. In this part, renewable generation is estimated as well as the bulk generations. From the total supply modeling point of view, two important generation types will be considered: 1) Bulk generations (most of the generations of this category are conventional, such as fossil fuel power plants), 2) renewables. The total generation of the first group can be calculated accurately. However, the generation estimation of renewables is not as achievable as bulk generations. In this cahpter, two influential renewables, wind and solar, are considered because of their considerable market share. The intermittency of wind and solar 1 generations helps us to achieve a better estimation of generation. Additionally, the uncertainty of these two renewable resources should be modeled based on two factors: 1) weather prediction (wind speed, solar radiation, and other weather parameters), 2) the performance of the solar panels and wind generators. This part leads to a generation model which will be utilized in the next simulations.

In order to validate the performance of the models in the proposed framework, an adequacy assessment is introduced in Section IV(C). Based on the forecasting model proposed, system adequacy can be estimated in terms of the system reliability.

Demand Forecasting

Assume that there are a set of customers C distributed over an area. By partitioning the area into n disjoint parts: $\{A_1, A_2, ..., A_n\}$, we obtain a corresponding partition of set $C = \{C_1, C_2, ..., C_n\}$ (communities of customers). The demand values of each subset C_q has been measured every μ units in time $[0, T\mu]$ for some integer T and real value μ. Then, assume that a year is divided into m parts (school time, Christmas holidays, summer break, etc.) based on the similarity of electricity usage pattern. We partition time interval $[0, T\mu]$ into m subsets: $\{I_1, I_2, ..., I_m\}$ such that I_i contains the ith part of every year belonging to $[0, T\mu]$. Additionally, every set I_i is divided to two parts: weekends I_{i1} and business days I_{i2}. Moreover, if a day is divided into d parts (again based on the similarity of electricity usage pattern during the day), we partition every interval I_{ij} into $\{I_{ij1}, I_{ij2}, ..., I_{ijd}\}$.

In addition, assume that we have the historical weather data in every area A_q over period $[0, T\mu]$. Considering that W denotes the set of different weather conditions, we partition time interval I_{ijk}, $\forall i = 1, ..., m$, $j = 1, 2$, $k = 1, ..., d$, in the following form for every area A_q, $\forall q = 1, ..., n$, $I_{ijk} = \bigcup_{w \in W} I_{ijk}^{(w,q)}$, where $I_{ijk}^{(w,q)}$ specifies the subset of I_{ijk} such that the weather condition in area A_q and time $t \in I_{ijk}^{(w,q)}$ is w.

Now, if interval $[0, T\mu]$ contains δ days, let D_v specifies the vth day of time interval $[0, T\mu]$ for every $v = 1, ..., \delta$. Additionally, consider $D(q, \tau)$ as the power demanded by the set of customers C_q measured at moment $\mu\tau$ (for every $\tau = 1, 2, ..., T$). For every interval $I_{ijk}^{(w,q,v)} = I_{ijk}^{(w,q)} \cap D_v$, if $I_{ijk}^{(w,q,v)} \neq \varnothing$, we specify five parameters: $X_1^{(q)}$ is the number of years passed since $t = 0$ (till interval $I_{ijk}^{(w,q,v)}$), $X_2^{(q)}$ is the number of weeks passed since the beginning of the ith partition of a year, $X_3^{(q)}$ is the number of days passed since the beginning of the jth partition of a week, $X_4^{(q)}$ is the temperature in area A_q and time interval $I_{ijk}^{(w,q,v)}$, and

$$y^{(q)} = \frac{\sum_{u\tau \in I_{ijk}^{(w,q,v)}} D(q, \tau)}{\sum_{u\tau \in I_{ijk}^{(w,q,v)}} 1}, \tag{1}$$

where $y^{(q)}$ specifies the average power demanded by the set of customers C_q in time interval $I_{ijk}^{(w,q,v)}$.

Maximum Likelihood Estimator. For every subset $I_{ijk}^{(w,q)} \subset [0, Tu]$, we construct a maximum-likelihood estimator for the dependent variable $y^{(q)}$ based on the following linear model:

$$\hat{y}^{(q)} = \left[1 \; X_1^{(q)} X_2^{(q)} X_3^{(q)} X_4^{(q)}\right]\left[\widehat{\beta_0} \, \widehat{\beta_1} \dots \widehat{\beta_4}\right]^T + N\left(0, \hat{\sigma}^2\right). \tag{2}$$

Considering that condition $I_{ijk}^{(w,q,v)} \neq \varnothing$ is only true for $v = v_1, v_2, \dots, v_p$, we obtain that: $Y^{(q)} = X^{(q)}\beta + \varepsilon$, $\forall q = 1, 2, \dots, n$, such that $Y^{(q)} = \left[y_1^{(q)} y_2^{(q)} \dots y_p^{(q)}\right]^T$, $\beta = \left[\widehat{\beta_0} \, \widehat{\beta_1} \dots \widehat{\beta_4}\right]^T$, $\varepsilon = \left[\varepsilon_1 \varepsilon_2 \dots \varepsilon_p\right]^T$, and

$$X^{(q)} = \begin{bmatrix} 1 & X_{11}^{(q)} & X_{12}^{(q)} & X_{13}^{(q)} & X_{14}^{(q)} \\ 1 & X_{21}^{(q)} & X_{22}^{(q)} & X_{23}^{(q)} & X_{24}^{(q)} \\ \vdots & \vdots & \vdots & \vdots & \vdots \\ 1 & X_{p1}^{(q)} & X_{p2}^{(q)} & X_{p3}^{(q)} & X_{p4}^{(q)} \end{bmatrix} \tag{3}$$

In Equation 1, $y_\iota^{(q)}$ specifies the average power demanded by the set of customers C_q in time interval $I_{ijk}^{(w,q,v)}$, moreover, $X_{\iota 1}, \dots, X_{\iota 4}$ denote the parameters on which $y_\iota^{(q)}$ is dependent, $\forall \iota = 1, 2, \dots, p$. Using the maximum-likelihood method for the linear model mentioned in Equation 2, we obtain that:

$$\hat{\beta}_{ML} = \left(X^{(q)T} X^{(q)}\right)^{-1} X^{(q)T} Y^{(q)T}, \tag{4}$$

$$\hat{\varepsilon}_{ML} = \left(Y^{(q)T} - X^{(q)}\hat{\beta}_{ML}\right) = N\left(0, \hat{\sigma}_{ML}^2 I\right), \tag{5}$$

$$\hat{\sigma}_{ML}^2 = \left(Y^{(q)T} - X^{(q)}\hat{\beta}_{ML}\right)\left(Y^{(q)T} - X^{(q)}\hat{\beta}_{ML}\right)^T / p. \tag{6}$$

Note that the ML estimator specified in Equation 2 can forecast the average power demand in an interval of few hours. However, by using the estimator repetitively and for different intervals $I_{ijk}^{(w,q)}$, we can forecast the average power demand for longer time; however, the variance of error will increase respectively. Additionally, the similar estimation model can be made for power generation. The only difference is that we don't need to partition a week into two parts. Moreover, we should partition a year into small parts based on the similarity of power generation pattern.

Short-Term Forecasting. In the previous subsection, we partitioned the interval $[0, Tu]$ into ~ $\left(\text{md}\,|W|\right)$ subsets in the form of $I_{ijk}^{(w,q)}$ (for every set of customers C_q). Additionally, for every subset $I_{ijk}^{(w,q)}$, a maximum likelihood estimator was constructed to estimate the average power demanded by customers C_q in time interval $I_{ijk}^{(w,q)} \cap D_v$. Our ultimate goal in this section is to construct an estimator for the value of power demanded by set of customers C_q in moment $t = \tau u$ (for some integer value τ) based on ARIMA(a,0,0) model with drift $-u^{(q)}$:

$$\left(1 - \sum_{l=1}^{a} \phi_l L^l\right)\left(D\left(q,\tau\right) - \mu^{(q)}\right) = \varepsilon_\tau,\tag{7}$$

where $\mu^{(q)}$ is the average of demand value $D\left(q,\tau\right)$ in time interval $t \in I_{ijk}^{(w,q,v)}$ which is estimated by Equation 2, ε_τ is a white noise of variance σ^2, $\tau u \in I_{ijk}^{(w,q,v)}$ for some $i,\ j,\ k, w,\ v$, and L is the lag operator: $L\left(D\left(q,\tau\right)\right) = D\left(q,\tau-1\right)$. By simplifying Equation 7, we obtain that:

$$D\left(q,\tau\right) = \left(\sum_{l=1}^{a} \phi_l - 1\right)\mu^{(q)} + \sum_{l=1}^{a} D\left(q,\tau-l\right) + \varepsilon_\tau.\tag{8}$$

By replacing $\mu^{(q)}$ with by $\hat{y}^{(q)} + \varepsilon'$ where $\varepsilon' = N\left(0, \hat{\sigma}_{ML}^2\right)$, we obtain that

$$
\begin{aligned}
D\left(q,\tau\right) &= \left(\sum_{l=1}^{a} \phi_l - 1\right)\left(\hat{y}^{(q)} + \varepsilon'\right) + \sum_{l=1}^{a} D\left(q,\tau-l\right) + \varepsilon_\tau \\
&= \left(\sum_{l=1}^{a} \phi_l - 1\right)\hat{y}^{(q)} + \sum_{l=1}^{a} D\left(q,\tau-l\right) + \varepsilon_\tau + \left(\sum_{l=1}^{a} \phi_l - 1\right)\varepsilon'
\end{aligned}\tag{9}
$$

Note that Equation 7 works only if random process $D\left(q,\tau\right)$ shows stationary behavior; otherwise, we need to use the model with moving average. In fact, assuming that process $D\left(q,\tau\right)$ is not stationary, ARIMA($a, 1, 0$) is much better for short-term forecasting:

$$\left(1 - \sum_{l=1}^{a} \phi_l L^l\right)(1 - L) D(q, \tau) = \varepsilon_\tau.$$

Consequently, we obtain that:

$$D(q, \tau) = (\phi_1 + 1)D(q, \tau - 1) + \sum_{l=2}^{a} (\phi_l - \phi_{l-1}) D(q, \tau - l) - \phi_a D(q, \tau - a - l) + \varepsilon_\tau$$

(10)

As can be seen, ARIMA($a, 1, 0$) model forecasts the demand value using its ($a + 1$) previous values with a white noise error. In addition, the power generation of the gth generator can also be forecast using ARIMA($a', 1, 0$). Assuming that $G(q, \tau)$ specifies the instantaneous power generated by the gth generator at moment t, we have: $\left(1 - \sum_{l=1}^{a'} \phi_l' L^l\right)(1 - L) G(q, \tau) = \varepsilon_\tau'$, or equivalently.

$$G(q, \tau) = (\phi_1' + 1)G(q, \tau - 1) + \sum_{l=2}^{a} (\phi_l - \phi_{l-1}) G(q, \tau - l) - \phi_{a'}' DG(q, \tau - a' - l) + \varepsilon_\tau'$$

(11)

In the following section, we analyze the adequacy of the electricity system based on ARIMA($a, 1, 0$) forecasting model.

Unit Commitment and Optimal Power Flow Problem

After long term demand and generation forecasting, we should solve the unit commitment (UC) problem. The main objective of solving UC problem is to calculate the most economical combination of all of the power generation units in order to meet estimated load (in the first stage) and required reserve for power system performance. Several constraints of generation units are considered in UC problem, such as minimum on/off time, ramping up/down, minimum/maximum generating capacity, and fuel and emission limit. The calculated power output of each generation unit will not meet the demand accurately and it needs more almost real-time scheduling to compensate the error of demand estimation. Therefore, we need the real-time generation estimation to ensure that generation meets load at each instant.

We define objective function F, of generating power from N units over a specific time horizon T. The total cost from each generator at a given period is the fuel cost, C_i, plus the any start-up costs, CS_i, that may be incurred during the period:

$$F = \sum_{t=1}^{T} \sum_{i=1}^{N} \left[C_i \left(P_i(t) \right) + S_i \left(x_i(t), u_i(t) \right) \right], \tag{12}$$

where the fuel costs, C_i, are dependent on the level of power generation $P_i(t)$. The start-up costs, S_i, are dependent on the state of the unit, x_i, which indicate the number of hours the unit has been on (positive) or off (negative), and the discrete decision variable, u_i, which denotes if power generation of the unit at time t is up (1) or down (-1) from the unit at time $t+1$. The objective function is to be minimized subject to a series of system and generator constraints. First, the system demand, $P_d(t)$, must be met $\sum_{i=1}^{N} P_i(t) = P_d(t)$.

There also must be sufficient spinning reserve, r_i, to ensure reliability. The required system spinning reserve is designated as P_r. The actual spinning reserve for a unit i is zero if the unit is off or $r_i = \min \left\{ P_i^{max}(t) - P_i(t), r_i^{max} \right\}$: i.e.,

$$\sum_{i=1}^{N} r_i(x_i(t), P_i(t)) \geq P_r(t).$$

In order to calculate the power flows in smart grid, with high penetration of distributed renewable resources, we should solve the power flow problem. The following equations show the power flow problem formulation and constraints.

minimize $\sum_{k} f_k \left(P_{G_k} \right)$ over P_G, Q_G, and V \hfill (13)

subject to:

$$
\begin{cases}
P_{G_k} - P_{D_k} = \sum_{i \in N} \Re \left\{ V_k \left(V_k^* - V_l^* \right) y_{kl}^* \right\} \\
Q_{G_k} - Q_{D_k} = \sum_{i \in N} \Im \left\{ V_k \left(V_k^* - V_l^* \right) y_{kl}^* \right\} \\
P_k^{min} \leq P_{G_k} \leq P_k^{max} \\
Q_k^{min} \leq Q_{G_k} \leq Q_k^{max} \\
V_k^{min} \leq \left| V_k \right| \leq V_k^{max} \\
\forall k
\end{cases}
$$

where N is the number of buses, $\left(P_{D_k} + jQ_{D_k} \right)$ is the load value at the kth bus, and $\left(P_{G_k} + jQ_{G_k} \right)$ is the generation value at the kth bus. Additionally, if there exists a line between the kth and l^{th} buses ($k, l \in N$), the following relations also constrain the mentioned minimization problem:

$$
\begin{cases}
\left| \theta_{kl} \right| = \left| \angle V_k - \angle V_l \right| \leq \theta_{kl}^{max} \\
\left| P_{kl} \right| = \left| \Re\{ V_k \left(V_k^* - V_l^* \right) y_{kl}^* \right| \leq P_{kl^{max}} \\
\left| S_{kl} \right| = \left| V_k \left(V_k^* - V_l^* \right) y_{kl}^* \right| \leq S_{kl^{max}} \\
\left| V_k - V_l \right| \leq V_{kl}^{max}
\end{cases}
\tag{14}
$$

The cost function f_k of generation units is as follow:

$$
f_i \left(p \right) = a + bp + cp^2
\tag{15}
$$

where a, b, and c are positive values. As the cost function of generation units is convex, we need potent convex optimization tools to solve this problem.

Adequacy Assessment

By assumption, we consider the maximum security for our electrical facilities (like cables, smart measurement equipment, etc.). Henceforth, the system reliability in our discussion refers to the system adequacy. In order to analyze the system adequacy, we need to use the forecasting models of instantaneous demand and generation presented in the previous section:

$$\begin{cases} D\big(q,\tau\big) = \hat{D}\big(q,\tau\big) + D_{\tau} \\ G\big(q,\tau\big) = \hat{G}\big(q,\tau\big) + G_{\tau} \end{cases} \tag{16}$$

$$\hat{D}\big(q,\tau\big) = (\phi_1 + 1)D\big(q,\tau - 1\big) + \sum_{l=2}^{a}\big(\phi_l - \phi_{l-1}\big)D\big(q,\tau - l\big) - \phi_a D\big(q,\tau - a - l\big). \tag{17}$$

$$\hat{G}\big(q,\tau\big) = \big(\phi_1' + 1\big)G\big(q,\tau - 1\big) + \sum_{l=2}^{a'}\big(\phi_l' - \phi_{l-1}'\big)G\big(q,\tau - l\big) - \phi_a' D\big(q,\tau - a' - l\big) \tag{18}$$

such that: and where D_{τ} and G_{τ} are two independent Gaussian white noises of the following covariance functions (regarding the Central-Limit theorem, the estimation errors of the instantaneous demand and generation are Gaussian processes):

$$\begin{cases} cov\big(s,t\big) = \sigma_g^2 \times \delta\big(s - t\big) \\ cov\big(s,t\big) = \sigma_d^2 \times \delta\big(s - t\big) \end{cases} \quad \forall s,t \geq 0$$

Now, assume that community C_q uses the qth renewable power plant (DRR) to satisfy its demand. Assuming that at given time t, community C_q has stored $S\big(q,\ t\big)$ units of energy, we obtain that:

$$S\big(q,t\big) = \int_0^t \big(G\big(q,t'\big) - D\big(q,t'\big)\big)dt' + s_q \quad \forall t \geq 0 \tag{19}$$

such that s_q is the initial stored energy in the community. By replacing the generation and demand functions with their equivalent random processes, we obtain that: where $\hat{S}\big(q,t\big) = \int_0^t \big(\hat{G}\big(q,t'\big) - \hat{D}\big(q,t'\big)\big)dt' + s_q$ and $W_t = \int_0^t \big(D_t - G_t\big)dt'$. Since D_{τ} and G_{τ} are two independent Gaussian white noises, W_t is a Wiener process of variance $(\sigma_g^2 + \delta_g^2)$. Here is the covariance function of process W_t:

$cov(W_s, W_t) = \min\{s, t\}(\sigma_g^2 + \delta_g^2)$. Moreover, $\hat{S}(q, t)$ is the expected value of the stored energy at given time t:

$$S(q, t) = \int_0^t \left(\hat{G}(q, t') - \hat{D}(q, t') + G_{t'} - D_{t'}\right)dt' + s_q$$

$$= \int_0^t \left(\hat{G}(q, t') - \hat{D}(q, t')\right)dt' + s_q - \int_0^t \left(D_{t'} - G_{t'}\right)dt' = \hat{S}(q, t) - W_t \tag{20}$$

- $\left[S(q, t)\right] = E\left[\hat{S}(q, t) - W_t\right] = \hat{S}(q, t) - E\left[W_t\right] = \hat{S}(q, t)$

According to the above analysis, the amount of stored energy $S(q, t)$ is equal to the summation of deterministic amount $\hat{S}(q, t)$ and the scaled Wiener process (W_t). In the rest of our analysis, we assume that the expected value of the stored energy never becomes less than the initial amount of energy (s_q); i.e. $\hat{S}(q, t) \geq s_q$, $\forall t \geq 0$. This condition can be held by providing sufficient DRRs for every community (which is designed based on long term forecasting of power demand and generation).

Here, we define the system adequacy ratio ($\rho_q(0, t)$) for the qth community as the probability that the actual stored energy $S(q, t')$ doesn't meet the low-threshold ($s_0 - \lambda$) for some $\lambda \in [0, s_q]$ and every $t' \in [0, t]$.

$$\rho_q(0, t) = \Pr\left[\forall t' \leq t : S(q, t') > s_0 - \lambda\right] = \Pr\left[\forall t' \leq t : \hat{S}(q, t') - W_{t'} > s_q - \lambda\right]$$
$$\geq \Pr\left[\forall t' \leq t : W_{t'} < \lambda\right] = \Pr\left[\sup_{t' \leq t}\{W_{t'}\} < \lambda\right] = \Pr\left[M_t < \lambda\right] \tag{21}$$

$$\rho_q(0, t) \geq \text{erf}\left(\frac{\lambda}{\sqrt{2t\sigma^2}}\right) \tag{22}$$

where M_t is the running maximum process corresponding to the scaled Wiener process W_t. So, regarding the characteristics of the running maximum process, we obtain the following inequality: such that $\sigma^2 = \sigma_g^2 + \delta_g^2$. In other words, we assume that if $S\left(q,t'\right) \le s_0 - \lambda$ (the stored energy becomes lower than some threshold in the qth community), the consumers demand will not be satisfied anymore.

If the DRR of each community is the only source of power for the community customers, the adequacy ratio will substantially decrease over time. In fact, even if the DRRs are sufficient to satisfy the customers' demand in long-term ($\hat{S}\left(q,t\right) \ge s_q$), the system adequacy cannot be guaranteed because of the white noise errors existed in the short-term forecasting scheme of demand and generation. Subsequently, we have to get help from the bulk generators located outside the community to cancel out the temporary noises and improve the adequacy ratio by generating extra energy on demand.

RECOMMENDED TESTBED FOR FUTURE APPLICATION OF FOG COMPUTING PLATFORM ON SMART GRID

To provide a context for validating the mechanisms of the Fog computing platform on smart grid, preliminary scenarios are carried out considering the sequence of failure illustrated in Figure 4 (a). In this figure, all data will be initially stored in the local storages. These data can be from distributed fault data to weather, load and irradiance data. Then command and control center will analyze the data, make the decisions, do the forecasting, perform the optimizations, and send the control signals. It has local and central sections. The IEEE 9 bus system can be used as the testbed in the proposed application, as shown in Figure 4 (b). Event 1 corresponds to a lightning strike on line 5 which leads to the operation of circuit breakers 13 and 14 and the reclosers which sense the over current. Event 2 will be a phase to phase short circuit fault on line 2.

Event 3 is a three phase to ground fault on line 3. Event3 is assumed to be an internal fault on Transformer 3 (e.g. insulation breakdown between winding and earth, transformer core fault, etc.) is occurred which results in failing of transmission lines 4 and 5. In Event 5 renewable generations (PV and wind) and/or related inverters are getting off while T3 is still faulty.

Event 1 and 2 are an asymmetric fault and Event 3 is a symmetrical fault which all three need sub-transient, transient and steady state analysis from the power system point of view. To check the clearance of these 3 events, reclosers will start operating. The first reclosing attempt will be made 2 seconds after the fault happening and 2nd and 3rd attempt will be made 15 seconds and 40 seconds after Lightning. Furthermore, fault ride through controllers on PV inverters and back to back converters of wind generation system will be enabled in all 5 cases. On the renewable side, LVRT controller on the inverters will be enabled to meet the interconnection standards.

Due to IEEE 1547a standard, if a voltage sag of 55% happens, it should be cleared in less than 1 second. And if the voltage sag is less than 40%, it can be cleared in less than 2 seconds. Moreover, the operation of other equipment like LTCs (Load Tap Changer), CBs (Capacitor Bank), VRs (Voltage Regulator), etc. will be studied. For Events 4 and 5 it is needed to enable the demand response and energy management control on top of all the other mentioned methods. Other important criteria which should be considered are the speed of data transmission to/from Fog storage and Fog computing center, speed of reading/writing data from/on Fog storage and speed of computational and control processes in comparison to the speed of local data transmission, local storage, and local computational an control processes.

FUTURE RESEARCH DIRECTIONS

Fog computing applications are one of the most useful techniques for the future smart grid development. Beyond the previously discussed applications for the fog, various future opportunities for fog -based energy management, information management and security in smart grid are discussed below with some of the research challenges (Bera, Misra, & Rodrigues, 2015).

- The flexibility of the smart grid is a very useful feature which makes it operate with or without real-time data and energy exchange. When a fault happens in the system smart grids should be able to give-and-take energy from/to the Fog energy storage devices.
- Virtual energy storing can be proposed to keep supply and demand curve all through peak hours. For this purpose, optimization (real time and online), the control flow of energy, and prediction will be done using Fog framework.
- Controller optimization for virtual power flow which will operate in both islanded mode and normal mode with any intermittencies of renewable power generation.
- Private and public Fog can be scaled and made interoperable to encourage global deployment in smart grid
- Scheduling of data traffic to reduce information traffic percentage in Fog environment and in the smart grid.
- Dynamic interaction of Fog services by customers to improve smart grid functionality through a decentralized power delivery.
- Fault protection algorithms will be implemented in a Fog-based structure in the future smart grid systems. In this framework, different kinds of equipment will be efficiently interacting with each other and multiple optimization algorithms will be implemented.

Figure 4. (a) Network base storage and computing framework. (b) Schematic view of the Smart Grid structure based on IEEE 9 bus system.

(a)

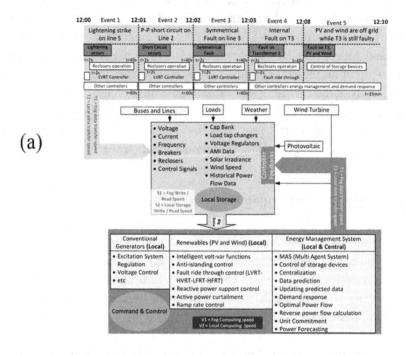

(b)

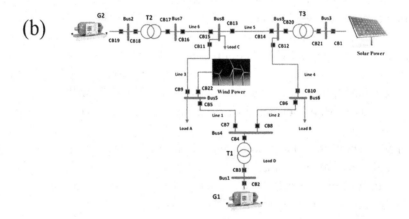

- Data leakage can happen intentionally or unintentionally and future smart grids are supposed to distinguish planned data leak by application of Fog computing technology. Nevertheless, accidental data leakage may happen occasionally in smart grids. Therefore, future grids need to differentiate the intentional from unintentional leakages and be able to avoid the latter.
- The state estimation, enterprising, and active management of the future smart grids will be done by Fog infrastructure. In state estimation speed is a crucial factor. As the rate of renewable integration into the grid is getting high, an instantaneous state estimation response will play an important role in the future and satisfactory policies can be defined by companies that provide power for a reliable energy distribution to the customers.

CONCLUSION

In this chapter, the significance of computing for smart grid was highlighted and justified in the context of three crucial scenarios: information management, energy management, and cyber-physical security. While the existing applications for processing and managing smart grid big data rely heavily on centralized frameworks such as the Fog, they will not prove scalable, interoperable, efficient and reliable under the future scenarios such as high levels of renewable integration into the transmission and distribution networks, and proliferation of large numbers of intelligent sensors across the grid. It is expected that network latency, congestion, and subsequent reduction in consumer QoS will make the Fog a stressed ecosystem for computing. This calls for a decentralized framework that distributes computational intelligence from the Fog towards the edge of the grid, called the Fog computing. Unlike the cloud, fog is capable of exhibiting scalability and interoperability with emerging future technologies for the smart grid. The effectiveness of the Fog for the three previously considered scenarios was discussed, following which the proposed Fog framework was introduced. Models for long/short-term forecasting, unit commitment, and adequacy assessment were elaborated. These models were then applied to a prototype fog ecosystem over a standard IEEE 9-bus testbed for validation.

ACKNOWLEDGMENT

This work is supported by the National Science Foundation under Grant No. 1553494. Any opinions, findings, and conclusions or recommendations expressed in this material are those of the authors and do not necessarily reflect the views of the National Science Foundation.

REFERENCES

Almadhor, A. (2016). A Fog Computing based Smart Grid Cloud Data Security. *International Journal of Applied Information Systems*, *10*(6), 1–6. doi:10.5120/ijais2016451515

Bera, S., Misra, S., & Rodrigues, J. J. P. C. (2015). Cloud Computing Applications for Smart Grid: A Survey. *IEEE Transactions on Parallel and Distributed Systems*, *26*(5), 1477–1494. doi:10.1109/TPDS.2014.2321378

Borcoci, E. (2016). Fog-computing versus SDN/NFV and Cloud computing in 5G. Paper presented in DataSys conference, Valencia, Spain.

Byers, C. C., & Wetterwald, P. (2015). *Fog Computing Distributing Data and Intelligence for Resiliency and Scale Necessary for IoT the Internet of Things*. Paper presented at ACM Symposium on Ubiquity.

Cisco. (2016). *Fog Computing and the Internet of Things: Extend the Cloud to Where the Things Are*. Retrieved from http://www.cisco.com/c/dam/en_us/solutions/trends/iot/docs/computing-overview.pdf

Dastjerdi, A.V., & Buyya, R. (2017, January). *Fog Computing: Helping the Internet of Things Realize its Potential*. IEEE Computer Society.

Deshmukh, U. A., & More, S. A. (2016). Fog Computing: New Approach in the World of Cloud Computing. *International Journal of Innovative Research in Computer and Communication Engineering*, *4*(9), 16310–16316.

Emfinger, W., Dubey, A., Volgyesi, P., Sallai, J., & Karsai, G. (2016). Demo Abstract: RIAPS – A Resilient Information Architecture Platform for Edge Computing. Paper presented at IEEE/ACM Symposium on Edge Computing (SEC), Washington, DC. 10.1109/SEC.2016.23

Fang, X., Misra, S., Xue, G., & Yang, D. (2012). Smart grid—the new and improved power grid: A survey. *IEEE Communications Surveys and Tutorials*, *14*(4), 944–980. doi:10.1109/SURV.2011.101911.00087

Guo, Y., Pan, M., & Fang, Y. (2012). Optimal power management of residential customers in the smart grid. *IEEE Transactions on Parallel and Distributed Systems*, *23*(9), 1593–1606. doi:10.1109/TPDS.2012.25

Gupta, H., Nath, S. B., Chakraborty, S., & Ghosh, S. K. (2016). *SDFog: A Software Defined Computing Architecture for QoS Aware Service Orchestration over Edge Devices*. arXiv preprint arXiv:1609.01190

Harp, D. R., & Gregory-Brown, B. (2013). *IT/OT Convergence: Bridging the Divide.* NexDefense ICS SANS Report. Retrieved from https://ics.sans.org/media/IT-OT-Convergence-NexDefense-Whitepaper.pdf

Khedkar, S. V., & Gawande, A. D. (2014). Data Partitioning Technique to Improve Cloud Data Storage Security. *International Journal of Computer Science and Information Technologies*, 5(3), 3347–3350.

Luntovskyy, A., & Spillner, J. (2017). *Smart Grid, Internet of Things and Fog Computing. Architectural Transformations in Network Services and Distributed Systems.* Wiesbaden, Germany: Springer. doi:10.1007/978-3-658-14842-3

Madsen, H., Burtschy, B., Albeanu, G., & Popentiu-Vladicescu, F. (2013). *Reliability in the utility computing era: Towards reliable Fog Computing.* Paper presented at 20th International Conference on Systems, Signals and Image Processing, Bucharest, Romania. 10.1109/IWSSIP.2013.6623445

National Electric Reliability Commission. (2012). *CIP-004-6: Cyber Security – Personnel & Training.* Retrieved from http://www.nerc.com/pa/Stand/Prjct2014XXCrtclInfraPrtctnVr5Rvns/CIP-004-6_CLEAN_06022014.pdf

National Electric Reliability Commission. (2014). *CIP-008-5: Cyber Security – Incident Reporting and Response Planning.* Retrieved from: http://www.nerc.com/pa/Stand/Project%20200806%20Cyber%20Security%20Order%20706%20DL/CIP-008-5_clean_4_(2012-1024-1218).pdf

National Electric Reliability Commission. (2014). *CIP-010-2: Cyber Security – Configuration Change Management and Vulnerability Assessments.* Retrieved from http://www.nerc.com/pa/Stand/Reliability%20Standards/CIP-010-2.pdf

National Electric Reliability Commission. (2014). *CIP-011-2: Cyber Security – Information Protection.* Retrieved from http://www.nerc.com/pa/Stand/Prjct2014XXCrtclInfraPrtctnVr5Rvns/CIP-011-2_CLEAN_06022014.pdf

Nelson, N., (2016). *The Impact of Dragonfly Malware on Industrial Control Systems.* SANS Institute InfoSec Reading Room Report.

Okay, F. Y., & Ozdemir, S. (2016) A Fog Computing Based Smart Grid Model. In *Proceedings of the International Symposium on Networks, Computers and Communications (ISNCC'16).* IEEE. 10.1109/ISNCC.2016.7746062

OpenFog Consortium Architecture Working Group. (2016). *OpenFog Architecture Overview.* Retrieved from https://www.openfogconsortium.org/wp-content/uploads/OpenFog-Architecture-Overview-WP-2-2016.pdf

OpenFog Consortium Architecture Working Group. (2017). *OpenFog Reference Architecture for Fog Computing*. Retrieved from https://www.openfogconsortium. org/wp-content/uploads/OpenFog_Reference_Architecture_2_09_17-FINAL.pdf

Popeanga, J. (2012). Cloud computing and smart grids. *Database Systems Journal, 3*(3), 57–66.

Pradhan, S., Dubey, A., Khare, S., Sun, F., Sallai, J., Gokhale, A., . . . Sturm, M. (2016). *Poster Abstract: A Distributed and Resilient Platform for City-Scale Smart Systems*. Paper presented at IEEE/ACM Symposium on Edge Computing (SEC), Washington, DC. 10.1109/SEC.2016.28

Rao, T.V.N., & Khan, M.A., Maschendra, M., & Kumar, M.K. (2015). A Paradigm Shift from Cloud to Fog Computing. *International Journal of Computer Science & Engineering Technology, 5*(11), 385–389.

Saharan, K. P., & Kumar, A. (2015). Fog in Comparison to Cloud: A Survey. *International Journal of Computers and Applications, 122*(3), 10–12. doi:10.5120/21679-4773

Sajjad, H. P., Danniswara, K., Al-Shishtawy, A., & Vlassov, V. (2016). *SpanEdge: Towards Unifying Stream Processing over Central and Near-the-Edge Data Centers*. Paper presented at IEEE/ACM Symposium on Edge Computing (SEC), Washington, DC. 10.1109/SEC.2016.17

Shi, W., & Dustdar, S. (2016, May 13). The Promise of Edge Computing. *IEEE Computer Society, 49*(5), 78–81. doi:10.1109/MC.2016.145

Sornalakshmi, K., & Vadivu, G. (2015). A Survey on Realtime Analytics Framework for Smart Grid Energy Management. *International Journal of Innovative Research in Science, Engineering and Technology, 4*(3), 1054–1058.

Ugale, B., Soni, P., Pema, T., & Patil, A. (2011). Role of cloud computing for smart grid of India and its cyber security. *Proc. IEEE Nirma Univ. Int. Conf. Eng.*, 1–5. 10.1109/NUiConE.2011.6153298

Wang, W., & Lu, Z. (2013). Cyber Security in the Smart Grid: Survey and Challenges. *Computer Networks, 57*(5), 1344–1371. doi:10.1016/j.comnet.2012.12.017

Yang, Y., Wu, L., & Hu, W. (2011). Security architecture and key technologies for power cloud computing. *Proc. IEEE Int. Conf. Transp., Mech., Electr. Eng.*, 1717–1720. 10.1109/TMEE.2011.6199543

Yanliang, W., Song, D., Wei-Min, L., Tao, Z., & Yong, Y. (2010). Research of electric power information security protection on cloud security. *Proceeding IEEE International Conference on Power System Technology (POWERCON)*, 1–6.

Yi, S., Hao, Z., Qin, Z., & Li, Q. (2015). *Fog Computing: Platform and Applications*. Paper presented at the Third IEEE Workshop on Hot Topics in Web Systems and Technologies. 10.1109/HotWeb.2015.22

Yi, S., Li, C., & Li, Q. (2015). *A Survey of Fog Computing: Concepts, Applications and Issues. ACM Mobidata ('15)*. Hangzhou, China: ACM. doi:10.1145/2757384.2757397

Zaballos, A., Vallejo, A., & Selga, J. (2011). Heterogeneous communication architecture for the smart grid. *IEEE Network*, *25*(5), 30–37. doi:10.1109/MNET.2011.6033033

Zhang, Y., Wang, L., Sun, W., Green, R. II, & Alam, M. (2011). Distributed intrusion detection system in a multi-layer network architecture of smart grids. *IEEE Transactions on Smart Grid*, *2*(4), 796–808. doi:10.1109/TSG.2011.2159818

Chapter 7
Mining Smart Meter Data:
Opportunities and Challenges

Ayushi Tandon
Indian Institute of Management Ahmedabad, India

ABSTRACT

Metering side of electricity distribution system has been one of the prime focus of industry and academia both. The most recent advancement in this field is installation of smart meters. The installation of smart meters enables collection of massive amounts of data regarding electricity generation and consumption. The analysis of this data could help generate actionable insights for the supply side and provide the consumers demand management-related inputs. The problem addressed in this chapter is to identify suitable data mining algorithm for applications like: estimating the demand and supply of electricity, user and use profiling of commercial, and industrial customers, and variables suitable for these purposes. This chapter, on the basis of rigorous literature review, presents a taxonomy of smart meter data mining. It includes the summary of application of smart meter data analytics, characteristics of dataset used, and smart meter business globally. This chapter could help researchers identify potential research opportunities, and practitioners can use it for planning and designing a smart electricity system.

DOI: 10.4018/978-1-5225-3996-4.ch007

This chapter published as an Open Access Chapter distributed under the terms of the Creative Commons Attribution License (http://creativecommons.org/licenses/by/4.0/) which permits unrestricted use, distribution, and production in any medium, provided the author of the original work and original publication source are properly credited.

INTRODUCTION

Electricity is vital utility as it supports many critical activities and services in human life. Electric power system consists of generation, transmission and distribution networks. The majority of electricity generation at present is centralized operation relying on non-renewable energy sources like coal, natural gas and nuclear. Electricity supply as an industry is highly capital intensive and suffers heavily due to demand-supply mismatch and transmission losses. In the recent decade application of Information and Communication Technology (ICT) in the power Industry has given rise to a new smart grid and smart metering infrastructure. Attempts are being made to enhance energy efficiencies and increase the reliability of electricity generation, transmission, and distribution systems by implementing network intelligence, empowering customers and enabling grid flexibility with smart devices deployments.

The evolution from the first known electricity meter, patented by Samuel Gardiner in 1872, towards a distributed electricity grid model able to manage numerous generation and storage devices in efficient and decentralized manner(Uribe-Pérez, Hernández, de la Vega, & Angulo, 2016). The metering side of the distribution system has been the focus, and the most recent advancement is the installation of *Smart Meters*. During 90's in the developed economies, automated meter reading (AMR) systems were introduced in the distribution networks moving away from electrochemical meters (Farhangi, 2010). Recently the AMR systems are being replaced with smart metering infrastructure (SMI), with two-way communication capabilities. These smart meters can digitally send meter reading to suppliers and also provide consumption and pricing related feedback to the user. This bidirectional communication ability of smart meters enables collection of the massive amount of data regarding the electricity generation and consumption. One of the biggest challenges in smart metering is the gathering and analysis of this 1.9 Terabytes of data per user/year. But, this also presents the most significant opportunity since analysis of this data could help generate actionable insights for supply side and provide the consumers demand management related inputs to the consumer. This opportunity if realized could enable countries to achieve the 20/20/20 European Sustainability Objectives of 20% reduction in energy consumption and, greenhouse gas emissions while 20% increase in renewables uses by 2020. According to the recent Benchmarking Report by the European Commission, the European Union has set the target to achieve 72% coverage of households with smart meters by 2020. In May 2017, Bangladesh Power Development Board (less developed country) launched smart meter service in the capital city, realizing the potential of this technology.

To realize these digital opportunities, all the companies part of electricity generation and distribution ecosystem need to transform their operations. To begin, they must develop a digital transformation strategy incorporating the understanding of smart meter data analytics. With the commercialization of smart meters, a huge amount of data is generated, which should be utilized for price and demand forecasting. However, the storage, querying, and mining of such smart meter data streams poses a significant challenge. The application of data mining algorithms on smart meter data could help meet both consumer's and power distribution utility's objectives. In this chapter, we present a rigorous review of academic literature and provide the taxonomy of application of smart meter data analytics. The researchers could refer this chapter to identify potential research opportunities and practitioners can use these findings for planning and designing an intelligent electricity system.

The guiding questions for this literature review are:

1. What algorithms are applied on smart meter data and for what purpose?
2. What are the characteristics of smart meter datasets and other external input parameters used for data mining?
3. What are the challenges associated with mining smart meter data and how such challenges are addressed in academic literature?

Most of the data analytics-based research is done to address either electricity demand or supply side issues in silos, so this literature review aggregate these results in a systematic manner.

LITTERATURE SEARCH

As the primary focus of this chapter is the application of data mining related algorithms on smart metering data, so we used search strings "smart meter + data mining," "smart grid+ data mining" and "smart meter +load forecasting." We restricted our search only to the articles having search string as part of the article title, keywords, and abstract. Also, we limited our review to articles published from 2008 till 2017. The search is conducted in the Scopus database along with Google Scholar and IEEE Xplore Digital Library. We did a manual scan of the article by reading abstract and rejecting irrelevant articles. Finally, we have summarized 35 articles as part of this chapter.

ORGANIZING FRAMEWORK

Intelligent smart metering interface (SMI) is one of the assorted technological innovations. This generates a large amount of data that has the potential of solving many challenges in power infrastructure. The main advantage of SMI is that it is bidirectional in nature and hence provides utility in the form of full visibility over consumption data and pricing. Many research articles have been published recently using SMI data. Data mining algorithms are primarily applied on this data for doing individual consumer profiling, load forecasting, tariff determination and differentiated load monitoring. We have organized the chapter under these broad headings.

- Consumer data analytics
- Dataset characteristics
- Privacy
- Algorithms applied
- Implementation examples

CONSUMER DATA ANALYTICS

Customer Segmentation and Load Forecasting

Customer segmentation help power distributors to identify interesting groups of customers. It is used to determine the tariff rates and to perform an optimal allocation of resources to meet electricity demand. Customers segmentation based on the consumption pattern is the useful exercise for power grids, because of overhead costs involved in scaling up and down of the generator. This segmentation also has the potential of inducing a change in consumption pattern by enforcing differential pricing based on average consumption, such that use pattern distributions are statistically similar within groups and statistically different across groups.

Alzate and Sinn (2013) applied Kernel Spectral Clustering algorithms on experimental time series data for residential customers and, small and medium enterprises. The model was trained to predict customer segment using auto regression. K-means clustering approach was used to group the customers based on similarities in their consumption behavior. Then, load forecasting was conducted for each group by applying multilayer perceptron neural network. Many researchers showed less error in prediction of energy requirements using neural networks and support vector machines (Savio, Karlik, & Karnouskos, 2010). Time variant characteristics of load data could also be used for power consumption pattern prediction (Zhang, Grijalva, & Reno, 2014). They proposed a layered tree structure using load, time and other attributes to determine the branch of the decision tree. These forecasting techniques could be used to identify customers having peak demands and high variability in power consumption. This group of customer is the potential target for demand response management offers. They should be sensitized to shift activities (like using a washing machine, dish washing) to the time of day when charges are low. (Quilumba, Lee, Huang, Wang, & Szabados, 2015).

For load forecasting context based mining of electricity use pattern by dynamically monitoring devices that operate together is also useful (Liao, Liao, Liu, Fan, & Omar, 2016). Sequential pattern mining i.e. identifying the set of activities that occur in sequence and require significant power consumption is an alternative to consumption pattern mining (Kouzelis & Bak-jensen, 2014). This paper suggested short-term load forecasting by finding frequently occurring patterns and representing the load data as a sequence of these states. Past sequences, similar to the present one, were sought to define a Markov-like process. The sequence prediction algorithm was used to determine rules, and Davies-Bouldin Indicator (DBI) was used to get the cluster of such sequences. Thus, probabilities for the future states of the load were predicted using present cluster information as Markov state. Certain data points in the dataset showed unusual consumption patterns, these subgroups were identified to improve classification accuracy, and their consumption patterns were analyzed separately.

Knowing Customer

Re-identification of customer using electricity consumption data from the smart meter is possible. The feature selection for re-identification is done by informed guessing like using prior knowledge regarding the schedule of customer or number of occupants etc. Research showed that household and consumer identification using smart meter data is possible with nearly 60% accuracy. Aman et al. (2011) conducted an experiment at the University campus and proposed a model for campus daily energy use prediction, which could be extended to city scale. The model utilized disaggregated data collected at appliance level, and this improved the forecasting accuracy of the model. Appliances were associated with well-defined finite ON/ OFF state and paper suggests that the timings of these events if detected could help in profiling of customers. The customer profiling like the number of members in the household, their age, education, etc. is useful for targeted advertising and creating energy use related awareness (Albert & Rajagopal, 2013). Fusco, Wurst, and Yoon (2012) suggested using power consumption data for the classification of households according to parameters like the presence of kids, ownership of specific appliances, employment status and education level of the residents. Many researcher articles are published on consumer profiling by using time series data of 24-hours power consumption (Beckel, Sadamori, & Santini, 2013; Lavin & Klabjan, 2015)

Beckel, Sadamori, and Santini (2014) also described how specific properties of a household – like its size or the number of individuals living in it – could be inferred from its electricity consumption profile. Profiling customer based on their electricity use pattern across day is done by applying techniques like K-means, K-medoid, and self-organizing maps (McLoughlin, Duffy, & Conlon, 2015). Mining smart meter could reveal consumer's income and education level (Viegas, Vieira, & Sousa, 2016). This technique could be applied to segment consumers based on their lifestyle like household activity time (Kwac, Flora, & Rajagopal, 2014). These results highlight need for research in the field of data security and anonymization of customer profile while transmitting smart meter data (Buchmann, Boehm, Burghardt, & Kessler, 2013).

Demand Management

The consumer of electricity is billed for its usage, but power generation is a complicated process with negative externality as it is heavily dependent on non-renewable fossil fuel. The main issue with power transmission and distribution is that the scaling up and down of the capacity is a costly process. Thus predicting demand and deciding the price is a challenge, which ICT-based smart power infrastructure is trying to address by applying data mining algorithms. Demand side management requires predicting the per-household energy consumption and informing same to the customer. This can be done easily and efficiently by using high-resolution data collected with the help of various sensors connected to appliances. Deployment of sensors in households provides high dimension data that can benefit standard electricity consumption prediction tools.

People get ready to leave for the office, school, etc. in the morning and get back at night in most parts of the world. The household consumption patterns show two peaks i.e. morning and evening time because of centralization of activities at these time. Marinescu, Harris, Dusparic, Clarke, and Cahill (2013), proposed a combination of several methods to predict better during these times and suggest to consumers a list of activities that can be shifted to some other time. This could peak off demand, by ensuring better load scheduling at the consumer end. To manage demand and influence customer behavior by generating useful insights the data needs to be stored efficiently and selectively. The predictive power of algorithm can be improved by doing variable selection that is an accurate representation of consumption pattern. The minimization of the number of the model input variable and forecasting error of model with the help of Genetic Algorithm- is suggested for simplified data-driven short-term forecasting of household electric loads (Niska, 2015).

Home appliance load modeling was done with the help of data collected using smart meters. The hidden markov model (HMM) was used to predict the state of appliances from aggregated power consumption data. Overall energy consumption due to the appliance of interest was calculated by using Gaussian probability distribution function in convention HMM (Guo, Wang, & Kashani, 2015). Article published in 2017 suggested using Gaussian Mixture Model for disaggregating household electricity consumption and identifying activities associated with appliances by approximating the distribution of power levels. The assumption in that paper was data points collected from meter were generated from a mixture of a finite number of Gaussian distributions and different modes of devices have different parameters associated with the Gaussian function (Cao, Wijaya, Aberer, & Nunes, 2017). These data mining and data visualization techniques could be applied for presenting each appliance related home energy consumption to the consumer in an easily comprehendible format. This way consumers would be encouraged to directly participate in demand response management which ultimately might lead to an overall energy consumption decrease.

Instead of using the multiple sensors to collect data, load disaggregating could be done from overall electricity consumption data to identify the start and stop time for most household appliances and certainty in their usage. The paper by Gajowniczek and Zabkowski applied C-means clustering and association rule mining to identify time-related patterns in appliance usage and the list of appliances used together (Gajowniczek & Zabkowski, 2015). Probability based standard kernel density estimation and conditional kernel density estimation (CKD) methods were applied to derive individual energy consumption function using previous week consumption data. This disaggregated data in combination with qualitative data collected from residents like: their perception towards dynamic pricing etc. could generate useful business insights. This information could be further utilized to provide meaningful feedback to the customers based on their energy time and appliance use profile (Stankovic, Stankovic, Liao, & Wilson, 2016).

This might help users in managing their demand during peak price hours as they will be informed charges associated with the use of the particular appliance by converting density estimates of consumption into density estimates of cost for different tariffs.

Suppliers could also use these demand estimates to devise innovative time-of-use pricing (Arora & Taylor, 2016). They could make data dependent decision making regarding the dynamic time of use pricing and this might encourage customers to shift their consumption resulting in significant savings. Time of use tariff is generally determined by clustering user group into three categories: commercial, industrial and residential. The tariff solution that flattens the demand curve is recommended because it is more efficient regarding energy generation (Zala & Abhyankar, 2014).

Distributors are also interested in integrating demand forecasts using smart meter data with renewable power generation data for efficient electricity distribution (Madureira, Gouveia, Moreira, Seca, & Lopes, 2013).

DATASET CHARACTERISTICS

With the commercialization of Smart meters, a huge amount of data regarding consumption pattern has been gathered. This is available in the form of the standard dataset for research purpose. Also, many sponsored projects are running at various universities, and data generated by such SMI is also used for research. The data could be logged using the smart meter at varying time intervals like every 15 minutes (most commonly used), every 30 minutes or hourly. The minimum sample size for data mining algorithms is 1000 data points.

Besides meter data, many exogenous factors like attributes from the university's academic calendar that indicate occupancy patterns, static knowledge of buildings such as surface area, and historical weather information were used as attributes in forecasting models by researchers (Aman et al., 2011). Research has shown that weather, seasonal variations, and house characteristics are the important determinant of house electricity consumption. Besides this building characteristics like stand alone or apartment, floor size of the building, the age of the residential colony is identified as factors that significantly affect household electricity consumption. Apartments equipped with energy-efficient lights and double-paned windows have significantly lower power consumption in comparison to households lacking both. House with pet consumes more energy and this consumption is high during summers because of air conditioner use to make home conformable for pets (Kavousian, Rajagopal, & Fischer, 2013). Thus data used for analysis needs to be identified by observing power consumption about time-of-day, the season of the year, house head activity, number of pets, etc.

Since forecasted weather impacts electricity usage hence Chakravorty, Rong, Evensen, & Wlodarczyk (2014) suggested its use to predict energy consumption along historical consumption data during similar weather conditions. They applied distributed unsupervised Gaussian-Means Clustering Algorithm in the paper. After including weather and pressure related data as predictors in energy consumption prediction models for buildings, the testing accuracy of four different algorithms was compared (Candanedo, Feldheim, & Deramaix, 2017). Selective load profiling i.e. identifying the relationship between outside air temperature and specific appliance electrical load in individual buildings for example AC electricity consumption based on external weather conditions is also useful (Dyson, Borgeson, Tabone, & Callaway, 2014). Jin et al. proposed a two-step method to forecast load at the household level by using local temperature, the day of week variable. Categorical variable indicating holiday is also used as the predictor in the model along with power consumption data (Jin et al., 2014). Many academic articles are published based on the assessment of the correlations between energy demand and the exogenous variables.

Hayes, Gruber, and Prodanovic, (2015) quantified the effect of variables which influence demand, at each level of aggregation into three broad categories: time-related (e.g. day, hour of day, and whether or not the day is a normal working day); historical (e.g. previous hour demand, previous week equivalent hour demand, previous 24 hour average); and weather-related (temperature has by far the greatest influence, but other weather factors such as humidity/precipitation, solar irradiation, and the wind can also have effects). External temperature and duration of the certain climatic condition like summer or winter influence the household electricity consumption. Also, some members and daily use device in the household is responsible for more energy consumption. Thus combining household characteristic along with temperature gradient could be used to predict power consumption and categorize customer. This could be used to recommend suitable demand management approach (Birt et al., 2012).

Dataset Related Challenges

As smart meter data is transferred digitally, noise is added to the signal. An important problem in load data analysis is therefore input validation, where the task is to distinguish between data corruption and a change in data pattern due to a random event or periodical patterns. The overall process of data validation, noise removal, and preparation for further analysis is often referred to as data cleansing.

Missing Values: The data might have many missing values that need to be estimated before using it for prediction. ARMA (autoregressive moving average) and ARIMA (autoregressive integrated moving average) are most commonly used for predicting missing data values. Chiky, Decreusefond, & Hebrail (2010) has suggested two approaches, (1) Based on the use of the past distribution of slopes in the time series; (2) A more sophisticated one based on a stochastic approach. These two methods are used for interpolating missing power consumption time series values and validated it using artificial dataset.

Data Storage: The data generated via SMI is huge and not all data attributes are used for analysis. Hence Data mining algorithms (like SOM) are used for data selection. Segmenting consumer is effective when besides using smart meter data, survey data like: weather, location, holidays, communities, etc. is also used (Wijaya, Ganu, Chakraborty, Aberer, & Seetharam, 2014). Thus smart meter data centers have to store appropriate consumption data along with other exogenous information variables available (Park, Ryu, Choi, Kim, & Kim, 2015). Only the segment identification information about consumers by clustering them based on load profile could be stored (Quilumba, Lee, Huang, Wang, & Szabados, 2014). This will save memory.

PRIVACY CONCERNS

There are many actors involved in smart meter ecosystem, and this adds to vulnerabilities of system design. The actors and factors part of this ecosystem are: customer, smart meter installation companies, smart meter reading staff, electricity generation units, electricity distribution companies (public and private entities), renewable energy devices at user end, climatic conditions, infrastructure of city, building layout and materials used, government policies and marketplace dynamics. External conditions (e.g., location and weather), physical characteristics of dwelling, appliance, and electronics stock, and occupants. The smart meter system design must be such that security concerns must be taken into account during each phase of the project, and even when the system becomes obsolete. The paper by Alabdulkarim and Lukszo deals in great detail about security requirements during system analysis, design, and implementation. It also recommends role base access of data collected to reduce malicious attack opportunities (Alabdulkarim & Lukszo, 2008).

Intrusion

Studies have shown that providing household's occupants with personalized breakdown of appliance energy consumption allows them to take steps towards reducing their total energy consumption. Putting sensors on each device and monitoring the sensor, could provide details about device activity. This is an intrusive method, and hence unsupervised training method for appliance load monitoring has been suggested (Parson, Ghosh, Weal, & Rogers, 2014). Researchers have developed methods for non-intrusive appliance signature extraction from turn ON and OFF events by detecting power spikes and clustering the suspect events (Dong, Meira, Xu, & Chung, 2013). Association rule mining technique could be used to prepare an electric signature database for various home appliances. This technique could be utilized for non-intrusive load monitoring (Dong et al., 2013). Power attributes like active power, reactive power, and harmonic content ranges, are appliance specific and hence helps locate appliance specific events. The on/off duration of an appliance is a critical variable to estimate the appliance's power consumption (Wang et al., 2012). Thus various methods are possible for load monitoring, and these methods have potential to reduce capital investment by the power supplier.

The commercialized smart meter's designs should be audited against hacking and attack by cyberterrorists. A detailed risk assessment should be carried out by the energy watchdog, and consumers should be informed to protect them from a range of fraudulent transactions for financial gain, remote disablement of critical device, etc.

ALGORITHMS USED

There are many algorithms and combination of algorithms in the academic literature for data mining. The researcher could do the comparative analysis of these algorithms for mentioned use case by varying the context and dataset characteristics like granularity of data captured. We have listed commonly used algorithm for the application mentioned above in the chapter as-

- **Data Pre-Processing:** Regression-based algorithms are most suitable for data cleansing and filling up missing values. Support vector machine, SVM is also used for data prediction.
- **Customer Segmentation:** Clustering is most commonly used for identifying customer segments. Spectral clustering, K-means clustering are commonly used as the base method and it shows performance improvement when heuristic based algorithms like genetic algorithm are applied to identify a number of clusters.

- **Load Disaggregation:** Probability density based methods are suitable for non-intrusive load monitoring. This application requires device energy usage and on/off state to be modeled using probabilistic functions like Gaussian function.
- **Feature Selection:** Artificial Neural Networks (ANN), and interview-based methods are most efficient in feature selection.

IMPLEMENTATION EXAMPLES

Texas, Austin is the hub of smart energy companies and university projects. Austin Energy, one of the largest electric utility companies in the US with more than 448,000 customer accounts. The company announced in October 2016 to install 2.5 lakh smart meters and these new meters will send information every 15 minutes, making it easier to diagnose outages.

Grid4C company a young startup, provides software products that make it possible to predict and optimize a smart electricity grid for both power producers, suppliers, and home customers.

Center- point Energy, a Houston-based utility company was awarded the 'International Smart Grid Action Network Award of Excellence 'and the 'Global Smart Grid Federation Best Smart Grid Project' for 2016. They use 50 days' historical data to determine how much wholesale supply is needed to serve the market. To integrate renewable off-grid energy production at consumer end CenterPoint planned the installation of new smart meters that could send power-off notifications to grid and signal when power demand was restored.

In Italy, Enel, the third- largest energy provider in Europe started deploying smart meters to about 27 million customers in 2015. The Enel's in-house analytics platform can optimize system-wide energy operations, including storage project, generation, transmission and distribution and micro grids.

In Korea, Korea Electric Power Corporation (KEPCO) started implementation of AMR based energy metering system for its industrial customers in 2000.

ElHub (Electricity Hub) will start operation in Norway, and its plan to facilitate the data exchange between market parties in Norway. ElHub aims at a standardization of data access to smart meter data for all eligible parties. This also presents research opportunity related to smart metering data storage and exchange format.

CONCLUSION

Application of information and communication technology (ICT) in the power industry has given rise to smart metering infrastructure (SMI). The smart meter ecosystem is a properly networked system which has two-way communication capabilities. The discussions about smart meter data management and data analytics in energy sector are driven by the smart meter roll-out and the need to enable customers to take energy saving decisions. Therefore, governments of most countries are at least discussing or have already implemented smart meter related policies. This in itself is expected to create a market of €6b+ for all IT vendors. The field offers varied research opportunities and many business organizations have been trying to extract meaningful and actionable insights from the large volumes of complex and high velocity SMI data streams.

Studies conducted on smart meter data show that smart metering system can help in peaking off the electricity demand and encouraging the use of off-grid renewable energy sources, by putting consumers in control of their energy use or by introducing time of use electricity tariff. With the commercialization of smart meters, a huge amount of standardized dataset is available easily, which can be utilized for price and demand forecasting. However, the storage, querying, and mining of such smart meter data streams poses significant challenge. The application of data mining algorithms on smart meter data could help meet both consumers' and power distribution utility's objectives. Consumers can also produce energy via various means such as solar and the wind, as well as distribute excess energy. However, due to the increasing share of renewables and the increasing number of new electricity devices (e.g. electric vehicles and battery storage) on the distribution grid level, power companies need to advance data analytics ability.

The smart meter has started transforming energy consumption as we know it and certainly brings plenty of exciting opportunities for customers and energy companies alike.

REFERENCES

Alabdulkarim, L., & Lukszo, Z. (2008). Information security assurance in critical infrastructures: Smart Metering case. *2008 1st International Conference on Infrastructure Systems and Services: Building Networks for a Brighter Future, INFRA 2008*. 10.1109/INFRA.2008.5439670

Albert, A., & Rajagopal, R. (2013). Smart meter driven segmentation: What your consumption says about you. *IEEE Transactions on Power Systems, 28*(4), 4019–4030. doi:10.1109/TPWRS.2013.2266122

Alzate, C., & Sinn, M. (2013). Improved electricity load forecasting via kernel spectral clustering of smart meters. *Proceedings - IEEE International Conference on Data Mining, ICDM*, 943–948. 10.1109/ICDM.2013.144

Aman, S., Simmhan, Y., & Prasanna, V. K. (2011). Improving energy use forecast for campus micro-grids using indirect indicators. *Proceedings - IEEE International Conference on Data Mining, ICDM*, 389–397. 10.1109/ICDMW.2011.95

Arora, S., & Taylor, J. W. (2016). Forecasting electricity smart meter data using conditional kernel density estimation. *OMEGA-International Journal of Management Science, 59*(A, SI), 47–59. 10.1016/j.omega.2014.08.008

Beckel, C., Sadamori, L., & Santini, S. (2013). Automatic socio-economic classification of households using electricity consumption data. *Journal of Economic Psychology, 3*(3–4), 75–86. doi:10.1016/0167-4870(83)90006-5

Beckel, C., Sadamori, L., Staake, T., & Santini, S. (2014). Revealing household characteristics from smart meter data. *Energy, 78*(SI), 397–410. 10.1016/j.energy.2014.10.025

Birt, B. J., Newsham, G. R., Beausoleil-Morrison, I., Armstrong, M. M., Saldanha, N., & Rowlands, I. H. (2012). Disaggregating categories of electrical energy end-use from whole-house hourly data. *Energy and Buildings, 50*, 93–102. doi:10.1016/j.enbuild.2012.03.025

Buchmann, E., Boehm, K., Burghardt, T., & Kessler, S. (2013). Re-identification of Smart Meter data. *Personal and Ubiquitous Computing, 17*(4, SI), 653–662. 10.100700779-012-0513-6

Candanedo, L. M., Feldheim, V., & Deramaix, D. (2017). Data driven prediction models of energy use of appliances in a low-energy house. *Energy and Buildings, 140*, 81–97. doi:10.1016/j.enbuild.2017.01.083

Cao, H.-A., Wijaya, T. K., Aberer, K., & Nunes, N. (2017). Estimating human interactions with electrical appliances for activity-based energy savings recommendations. In *Proceedings - 2016 IEEE International Conference on Big Data, Big Data 2016* (pp. 1301–1308). IEEE. 10.1109/BigData.2016.7840734

Chakravorty, A., Rong, C., Evensen, P., & Wlodarczyk, T. W. (2014). A distributed gaussian-means clustering algorithm for forecasting domestic energy usage. *Proceedings of 2014 International Conference on Smart Computing, SMARTCOMP 2014*, 229–236. 10.1109/SMARTCOMP.2014.7043863

Chiky, R., Decreusefond, L., & Hébrail, G. (2010). Aggregation of asynchronous electric power consumption time series knowing the integral. In *Advances in Database Technology - EDBT 2010 - 13th International Conference on Extending Database Technology, Proceedings* (pp. 663–668). Academic Press. 10.1145/1739041.1739122

Dong, M., Meira, P. C. M., Xu, W., & Chung, C. Y. (2013). Non-intrusive signature extraction for major residential loads. *IEEE Transactions on Smart Grid, 4*(3), 1421–1430. doi:10.1109/TSG.2013.2245926

Dyson, M. E. H., Borgeson, S. D., Tabone, M. D., & Callaway, D. S. (2014). Using smart meter data to estimate demand response potential, with application to solar energy integration. *Energy Policy, 73*, 607–619. doi:10.1016/j.enpol.2014.05.053

Farhangi, H. (2010). The Path of the Smart Grid 18. *IEEE Power & Energy Magazine, 8*, 18–28. doi:10.1109/MPE.2009.934876

Fusco, F., Wurst, M., & Yoon, W. J. (2012). Mining Residential Household Information from Low-resolution Smart Meter Data. *21st International Conference on Pattern Recognition (ICPR 2012)*, 3545–3548.

Gajowniczek, K., & Zabkowski, T. (2015). Data mining techniques for detecting household characteristics based on smart meter data. *Energies*, *8*(7), 7407–7427. doi:10.3390/en8077407

Guo, Z., Wang, Z. J., & Kashani, A. (2015). Home Appliance Load Modeling From Aggregated Smart Meter Data. *IEEE Transactions on Power Systems*, *30*(1), 254–262. doi:10.1109/TPWRS.2014.2327041

Hayes, B., Gruber, J., & Prodanovic, M. (2015). Short-Term Load Forecasting at the local level using smart meter data. *IEEE Eindhoven PowerTech*, 1–6. 10.1109/PTC.2015.7232358

Jin, N., Flach, P., Wilcox, T., Sellman, R., Thumim, J., & Knobbe, A. (2014). Subgroup discovery in smart electricity meter data. *IEEE Transactions on Industrial Informatics*, *10*(2), 1327–1336. doi:10.1109/TII.2014.2311968

Kavousian, A., Rajagopal, R., & Fischer, M. (2013). Determinants of residential electricity consumption: Using smart meter data to examine the effect of climate, building characteristics, appliance stock, and occupants' behavior. *Energy*, *55*, 184–194. doi:10.1016/j.energy.2013.03.086

Kouzelis, K., & Bak-jensen, B. (2014). *A Simplified Short Term Load Forecasting Method Based on Sequential Patterns*. Academic Press.

Kwac, J., Flora, J., & Rajagopal, R. (2014). Household Energy Consumption Segmentation Using Hourly Data. *IEEE Transactions on Smart Grid*, *5*(1), 420–430. doi:10.1109/TSG.2013.2278477

Lavin, A., & Klabjan, D. (2015). Clustering time-series energy data from smart meters. *Energy Efficiency*, *8*(4), 681–689. doi:10.100712053-014-9316-0

Liao, Y.-S., Liao, H.-Y., Liu, D.-R., Fan, W.-T., & Omar, H. (2016). Intelligent Power Resource Allocation by Context-Based Usage Mining. In *Proceedings - 2015 IIAI 4th International Congress on Advanced Applied Informatics, IIAI-AAI 2015* (pp. 546–550). IIAI. 10.1109/IIAI-AAI.2015.165

Madureira, A., Gouveia, C., Moreira, C., Seca, L., & Lopes, J. P. (2013). Coordinated management of distributed energy resources in electrical distribution systems. *2013 IEEE PES Conference on Innovative Smart Grid Technologies, ISGT LA 2013*. 10.1109/ISGT-LA.2013.6554446

Marinescu, A., Harris, C., Dusparic, I., Clarke, S., & Cahill, V. (2013). Residential electrical demand forecasting in very small scale: An evaluation of forecasting methods. *2013 2nd International Workshop on Software Engineering Challenges for the Smart Grid, SE4SG 2013 - Proceedings*, 25–32. 10.1109/SE4SG.2013.6596108

McLoughlin, F., Duffy, A., & Conlon, M. (2015). A clustering approach to domestic electricity load profile characterisation using smart metering data. *Applied Energy*, *141*, 190–199. doi:10.1016/j.apenergy.2014.12.039

Niska, H. (2015). *Evolving Smart Meter Data Driven Model for Short- Term Forecasting of Electric Loads*. Academic Press.

Park, S., Ryu, S., Choi, Y., Kim, J., & Kim, H. (2015). Data-driven baseline estimation of residential buildings for demand response. *Energies*, *8*(9), 10239–10259. doi:10.3390/en80910239

Parson, O., Ghosh, S., Weal, M., & Rogers, A. (2014). An unsupervised training method for non-intrusive appliance load monitoring. *Artificial Intelligence*, *217*, 1–19. doi:10.1016/j.artint.2014.07.010

Quilumba, F. L., Lee, W. J., Huang, H., Wang, D. Y., & Szabados, R. (2014). An overview of AMI data preprocessing to enhance the performance of load forecasting. *2014 IEEE Industry Application Society Annual Meeting, IAS 2014*, 1–7. 10.1109/IAS.2014.6978369

Quilumba, F. L., Lee, W.-J., Huang, H., Wang, D. Y., & Szabados, R. L. (2015). Using Smart Meter Data to Improve the Accuracy of Intraday Load Forecasting Considering Customer Behavior Similarities. *IEEE Transactions on Smart Grid*, *6*(2), 911–918. doi:10.1109/TSG.2014.2364233

Savio, D., Karlik, L., & Karnouskos, S. (2010). Predicting energy measurements of service-enabled devices in the future smartgrid. *UKSim2010 - UKSim 12th International Conference on Computer Modelling and Simulation*, 450–455. 10.1109/UKSIM.2010.89

Stankovic, L., Stankovic, V., Liao, J., & Wilson, C. (2016). Measuring the energy intensity of domestic activities from smart meter data. *Applied Energy*, *183*, 1565–1580. doi:10.1016/j.apenergy.2016.09.087

Uribe-Pérez, N., Hernández, L., de la Vega, D., & Angulo, I. (2016). State of the Art and Trends Review of Smart Metering in Electricity Grids. *Applied Sciences*, *6*(3), 68. doi:10.3390/app6030068

Viegas, J. L., Vieira, S. M., & Sousa, J. M. C. (2016). Mining consumer characteristics from smart metering data through fuzzy modelling. *Communications in Computer and Information Science*, *610*, 562–573. doi:10.1007/978-3-319-40596-4_47

Wang, Y., Hao, X., Song, L., Wu, C., Wang, Y., Hu, C., & Yu, L. (2012). Tracking states of massive electrical appliances by lightweight metering and sequence decoding(重点). *Proceedings of the Sixth International Workshop on Knowledge Discovery from Sensor Data*, 34–42. 10.1145/2350182.2350186

Wijaya, T. K. b, Ganu, T., Chakraborty, D., Aberer, K., & Seetharam, D. P. (2014). Consumer segmentation and knowledge extraction from smart meter and survey data. In *SIAM International Conference on Data Mining 2014, SDM 2014* (Vol. 1, pp. 226–234). SIAM. 10.1137/1.9781611973440.26

Zala, H. N., & Abhyankar, R. (2014). A novel approach to design time of use tariff using load profiling and decomposition. *2014 IEEE International Conference on Power Electronics, Drives and Energy Systems (PEDES)*, 1–6. 10.1109/PEDES.2014.7042027

Zhang, X., Grijalva, S., & Reno, M. J. (2014). A time-variant load model based on smart meter data mining. *IEEE PES General Meeting / Conference & Exposition*, 1–5.

KEY TERMS AND DEFINITIONS

Data Mining: It is practice and process of identifying patterns in large datasets with the help of computing algorithms.

Feature Selection: Identifying attributes or variables relevant for particular prediction model is called as feature selection or attribute selection or variable selection.

K-Means Clustering: This is unsupervised process of partitioning data into k-clusters, so as to minimize the distance between point and nearest cluster mean.

Probability Density Function: A function of a continuous random variable, whose integral across an interval gives the probability that the value of the variable lies within the same interval.

Self-Organizing Maps: It is a data visualization technique which helps to understand high dimensional data by reducing the dimensions of data to a map.

Smart Meter: It is advance meter that displays real time information about electricity consumption and costing to the consumer. It is capable of sending and receiving information using radio waves, like mobile phones and TVs.

Support Vector Machine: It is used to categories new unseen data points into separate groups based on their properties and a set of known examples, which are already categorized. It works by representing the features with the help of finite dimension vector space.

Chapter 8
Communications Technologies for Smart Grid Applications:
A Review of Advances and Challenges

Gurkan Tuna
Trakya University, Turkey

Resul Daş
Fırat University, Turkey

Vehbi Cagri Gungor
Abdullah Gul University, Turkey

ABSTRACT

Smart grid is a modern power grid infrastructure for improved efficiency, reliability, and safety, with smooth integration of renewable and alternative energy sources, through automated control and modern communications technologies. The smart grid offers several advantages over traditional power grids such as reduced operational costs and opening new markets to utility providers, direct communication with customer premises through advanced metering infrastructure, self-healing in case of power drops or outage, providing security against several types of attacks, and preserving power quality by increasing link quality. Typically, a heterogeneous set of networking technologies is found in the smart grid. In this chapter, smart grid communications technologies along with their advantages and disadvantages are explained. Moreover, research challenges and open research issues are provided.

DOI: 10.4018/978-1-5225-3996-4.ch008

Copyright © 2018, IGI Global. Copying or distributing in print or electronic forms without written permission of IGI Global is prohibited.

INTRODUCTION

Electricity is the widely used form of energy and modern society significantly depends on it. On the other hand, although the global demand for electricity is growing tremendously, most of the electrical power systems were built up over more than a century. In addition, most of the installed electricity generation capacity relies heavily on fossil fuels, and this increases carbon dioxide in the atmosphere and makes a significant contribution to climate change.

To fulfill the increasing demand for power and mitigate the consequences of climate change, an electric system that can address these challenges in an economic, sustainable and reliable way is needed (Gungor et al., 2013). In this respect, smart grids can meet the rising electricity demand, enhance energy efficiency, increase quality and reliability of power supplies, and integrate low carbon energy sources into power networks as shown in Figure 1 (Liserre, Sauter, & Hung, 2010). Basically, a smart grid is an evolved power grid built on sophisticated infrastructure that manages electricity demand in an economic, reliable and sustainable manner.

Different from the traditional power grids, smart grids apply high standards to the control capability of all facilities. Because, although in the traditional power grids, power flows travelled from generators to customers, in smart grids, due to the increasing use of distributed power generation relying on renewable resources which fluctuate over time, there are additional sources of power flows which an electric utility must contend with (Lu, Kanchev, Colas, Lazarov, & Francois, 2011;

Figure 1. Smart Grid

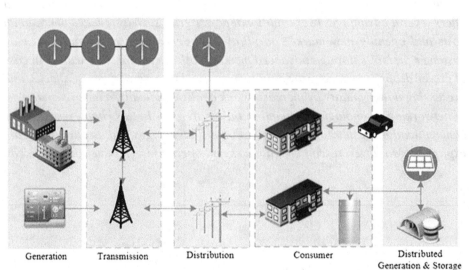

| Generation | Transmission | Distribution | Consumer | Distributed Generation & Storage |

Saber, & Venayagamoorthy, 2011). The distributed power generated by small units also requires a high degree of flexibility during distribution. In favour of various environmental and economic factors, in addition to large-scale integration of renewable energy sources, smart grids accommodate Demand Response (DR) capacity to help balance electrical consumption with supply (Palensky & Dietrich, 2011), as well as the potential for integrating new technologies to utilise energy storage devices and enable efficient use of plug-in electric vehicles.

Since in smart grids the ability to control the electricity supply gives consumers new ways to influence demand, accurate models of user behaviour are needed to assess control algorithms which realise admission and policing of flows from utility-owned or consumer-owned power sources. In this regard, there is an obvious need for communications networks which support reliable and secure information transfer between the various entities in the evolved power grids. However, although there are already modern communications technologies which can provide a secure and reliable communications system between the production facilities of electricity networks, electricity consumers and devices, in terms of network performance, suitability, interoperability and security, there are many issues that need to be resolved (Gungor et al., 2011).

It has been already seen that there is no silver bullet for building smart grid communications networks and although the desire for all Wide Area Network (WAN), Neighbourhood Area Network (NAN) and Home Area Network (HAN) applications is the highest available bandwidth and lowest latency, it is not possible in most scenarios due to cost-related reasons and the most cost-effective communications networks consisting of numerous platforms which serve the varying latency, bandwidth, and prioritization needs of different kinds of smart grid applications are preferred by electric utilities (Kuzlu, Pipattanasomporn, & Rahman, 2014). Moreover, compared to the past, even if high performance communications networks are available at lower costs, strategies for integrating, operating, and managing various communications technologies and vendor solutions should be developed in order to optimize smart grid investments. In this chapter, a survey on communications technologies for smart grid applications is given. The chapter mainly focuses on the identification of opportunities to accommodate communications protocols which have been designed for traffic control to enable quality of service (QoS) for smart grid applications and to manage power flows between traditional and renewable generation sources and between utility-owned and consumer-owned assets in smart grids. Moreover, in this chapter, research challenges and open research issues are presented. The rest of the chapter is structured as follows. Section 2 reviews existing and emerging smart grid applications. A detailed review of the communications technologies and standards is presented in Section 3. Research challenges and future research directions are given in Section 4 and Section 5, respectively. Finally the chapter is concluded in Section 6.

SMART GRID APPLICATIONS

In recent years, electrical systems have started to undergo a major evolution sometimes called "smart grid transformation" in order to improve reliability and reduce electrical losses, capital expenditures and maintenance costs. In some smart grid implementations, High-Voltage Direct Current (HVDC) transmission systems are preferred. In contrast with the common alternating current systems, HVDC transmission systems use direct current for the bulk transmission of electrical power (Li et al., 2013). Hence, for long-distance transmissions, they can be less expensive and suffer lower electrical losses.

In a smart grid, endpoints located on premise and along the power line can interact with each other. Common interactions carried out in a smart grid rely on a set of commands to connect, to disconnect, to check status, to turn on, to turn off and to read data. Compared to the tradional power grids, smart grids provide not only a more reliable energy supply for their consumers but also higher control over energy costs. In addition, smart grids include other benefits such as integration of more renewable power sources, reduced peak demand, and reduced CO_2 emissions and other pollutants (Saber & Venayagamoorthy, 2011).

In a smart grid, communications technologies are responsible for relaying information from the premises of usage to electric utility and contrariwise. Basic operations of a smart grid rely on two primary smart grid applications: Automatic Meter Reading (AMR) and Supervisory Control and Data Acquisition (SCADA) systems/Distribution Automation Systems (DASs) (Mahmood, Aamir, & Anis, 2008). Initially, AMR systems relied on proprietary, low-power wireless systems to support drive-by and walk-up meter data collection. Similarly, SCADA systems/DASs initially used proprietary wireless systems to support telemetry and control functionality. However, later on, AMR systems started to support full remote meter reading using fixed networks which supported telemetry without requiring proximity. These AMR networks used proprietary wireless, mesh or point-to-point, networks or proprietary Power Line Communications (PLC) technologies (Bumiller, Lampe, & Hrasnica, 2010; Yigit, Gungor, Tuna, Rangoussi, & Fadel, 2014). Although, early, proprietary fixed-network AMR systems supported very low data rates sufficient only for meter reading and a few simple operations, electric utilities broadened their business cases beyond AMR to include applications such as outage/restoration and price signalling, and they began considering deployment of Advanced Metering Infrastructure (AMI).

While AMR systems gather diagnostic and consumption data from smart meters to analyze and troubleshoot and accurately bill consumers, DASs aid in efficient operations of a reliable power line. Basically, DASs monitor a particular region with transmission and distribution (T&D)-level field equipments and through automated decision-making, quick fault detection and efficient power restoration capabilities, enable electric utilities to measure voltage and current levels and control assets in their networks (Gomes, Colunga, Gupta, & Balasubramanian, 2015). Although basic forms of smart grid implementations which rely on those primary systems are not obsolete today, the industry moved to more advanced functionalities that rely on two-way communication and real time capability. By enhancing existing AMR systems with two way communication and real time capability, a new system called AMI has been created (Hart, 2008). AMI is a term that denotes automated two-way communication infrastructure between smart meters and the data center of an electric utility. Similarly, in parallel with the advancements, other smart grid applications have been developed and implemented. In the following major smart grid applications and their roles are briefly reviewed, and a summary of communications technologies for common smart grid applications are listed in Table 1. In the table, since the security and reliability requirements of all smart grid applications are high due to the role of smart grids, these fields are omitted.

- **AMI:** Since other utility network components of a smart grid will be structurally covered with AMI technologies, it is basically a logical starting point to review smart grid communications technologies. AMI comprises of four main tiers, namely Home Area Network (HAN), smart meter, concentration point and utility data center.
- **Meter Data Management System (MDMS):** It allows automating meter data collection on a secure network, tracking meter data from advanced utility meters in a central database, and producing accessible energy reports.
- **DR:** It is a term used for programs designed to encourage consumers to make short-term reductions in energy demand in response to a price signal from the electricity hourly market, or a trigger initiated by the grid operator (Gungor et al., 2013). Usually, DR actions are in the range of 1 to 4 hours and include turning off or dimming banks of lighting, adjusting Heating, Ventilating and Air Conditioning (HVAC) levels, or shutting down a portion of a manufacturing process. Onsite generation can be used to displace load drawn from the power grid, as well.

- **Demand Side Management (DSM):** They are programs used to encourage consumers to be more energy efficient. DSM measures typically include lighting retrofits, building automation upgrades, re-commissioning, HVAC improvements, and variable frequency drives (Palensky, & Dietrich, 2011).
- **Advanced Distribution Management System (ADMS):** Along with real-time visualisation and monitoring of network status, it provides a host of analytical tools which recommend the most optimal device operations to maximise network efficiency and reliability (Meliopoulos, Polymeneas, Tan, Huang, & Zhao, 2013). Its network simulation capability helps forecast medium-term and long-term load and supports reliable development and planning.
- **Wide Area Situational Awareness (WASA):** WASA makes use of advanced measurement technologies in order to monitor and control large power grids and super grids and can be used as a standalone infrastructure or by complementing other conventional supervision systems. Phasor Measurement Units (PMUs) are the initial data source for WASA applications, PMUs are devices that provide high quality measurements of bus angles and frequencies using a common time source for synchronisation such as GPS radio clock (Matsumoto et al., 2012). PMUs can be autonomous systems or part of a protective relay or other device in a substation and are capable of detecting faults early, increasing the power quality, enabling load shedding and other load control techniquesessential in regional transmission grids, local distribution grids and even wide super grids.
- **Transmission Line Monitoring System (TLMS):** It provides continuous data on conductor behaviour required for facility rating, regulatory compliance, and Dynamic Line Rating (DLR). DLR uses sensors to identify the current carrying capability of a section of the network in real-time to optimise utilisation of existing transmission assets, without the risk of causing overloads (Giannuzzi, Pisani, Vaccaro, & Villacci, 2015).
- **Substation Automation System (SAS):** It provides solutions for the reliable control, protection, operation and monitoring of electricity substations.
- **Asset Management System (AMS):** It is for improving the utilization of T&D assets and more effectively managing these assets' life cycle. It requires smart sensors to provide operational and asset condition information to significantly improve asset management.
- **Home Energy Management System (HEMS):** A combined HAN with hardware which has auto-pricing response capabilities, DR load control, and home automation controls.

- **Vehicle-to-Grid (V2G):** It presents a mechanism to meet key requirements of the electric power system, using electric vehicles when they are parked and underutilized. An electric vehicle can be used as both a load and a generating source to balance the system frequency by charging the battery when there is too much generation in the power grid and acting as a generator by discharging the battery when there is too much load in the power grid system (Su, Eichi, Zeng, & Chow, 2012). In addition to this, electric vehicles can provide other services including: spinning reserve, back-up service, and peak management.
- **Distributed Energy Resources and Storage (DERS):** It consists of small-scale renewable energy sources and energy storage to balance load and capacity without building large-scale generation.
- **Outage Management System (OMS):** It is a utility network management software application to model network topology for efficient and safe field operations related to outage restoration (Gungor et al., 2013). Tightly integrating with call centers, it provides timely, accurate, customer-specific outage information. In addition, it integrates with SCADA systems for real-time-confirmed switching and breaker operations.

Table 1. Communications requirements for common smart grid applications

Smart Grid Application	Bandwidth (min)	Latency (max)
AMR	10 Kbps	10 sec
AMI	10 Kbps (per node), 500 Kbps (backhaul)	2 sec
DR	14 Kbps (per node)	A few min
SCADA	10 Kbps	4 sec
WASA (Synchrophasor)	600 Kbps	200 msec
SAS	9.6 Kbps	200 msec
AMS	56 Kbps	2 sec
OMS	56 Kbps	2 sec
TLMS	9.6 Kbps	200 msec
HEMS	9.6 Kbps	2 sec
V2G	9.6 Kbps	5 min
DERS	9.6 Kbps	2 sec

(Department of Energy, 2010; Gungor et al., 2011; Kuzlu, Pipattanasomporn, & Rahman, 2014; Yang, Barria, & Green, 2011).

SMART GRID COMMUNICATIONS TECHNOLOGIES

To enable sensing, monitoring, communication and control capabilities, smart grid communications infrastructure must address a number of key issues including performance in terms of latency, data throughput and reliability, scalability, security, availability, costs, control and technology life-cycle (Wenpeng, Sharp, & Lancashire, 2010). However, most electric utilities are aware of the fact that their communications networks are not sufficient enough to support their smart grid activities over the next few years due to the completion of their current AMI deployments, the growth of new applications in DASs and the increasing adoption of distributed generation and electric vehicles (Fan et al., 2013). Hence, they are planning to deploy many or more devices in their networks, possibly more intelligent endpoints, and upgrade their networks. Since the growth in the number of endpoints causes a great increase in the volume of data transmitted for utility operations, higher-bandwidth and lower-latency networks are needed to address the increased supply of data (Garcia-Hernandez, 2011).

Although wired communications technologies such as optical fibre and Digital Subscriber Line (DSL) technologies provide some of the most reliable, high-performance networks, the high cost of deployment is a barrier in many cases. On the contrary, lower-performance technologies including telephone, copper and coaxial cable, and satellite are not preferred because of the high bandwidth and low latency demands of new smart grid applications. Therefore, most utilities prefer wireless communications technologies such as WiMAX (Worldwide Interoperability for Microwave Access), Wi-Fi (Wireless Fidelity), cellular and licensed RF for AMI and DAS. Cost, reliability, performance, technology longevity, deployment region and existing utility communications equipment play the key roles in decision-making pertaining to communications networks during the next several years, with each technology requiring the utility to make some tradeoffs (Gungor et al., 2011). Table 2 lists a summary of typical communications technologies for common smart grid applications, and the following subsections review existing and emerging communications technologies for HAN, NAN and WAN applications.

Since smart grids considerably depend on communications in order to coordinate the generation, distribution, and consumption of energy and there is a variety of communications partners, a heterogeneous two-tier infrastructure which consists of IP-based and suitable field-level networks with particular attention to metering and supervisory control and data acquisition applications may be the most appropriate solution (Sauter, & Lobashov, 2011; Meiling, Schmidt, & Steinbach, 2015; Zaballos, Vallejo, & Selga, 2011).

Table 2. Typical communications technologies for common smart grid applications

Technology	Specifications (Spectrum / Maximum Data Rate / Maximum Coverage-Range)	Advantages	Disadvantages	Smart Grid Networks / Applications
GSM	900 MHz, 1800 MHz / 14.4 Kbps / 1-10 km	Worldwide availability	Low data rate	HAN, AMI, DR
GPRS	900 MHz, 1800 MHz / 170 Kbps / 1-10 km	Worldwide availability	Low data rate	HAN, AMI, DR
3G	1.92-1.98 GHz, 2.11-2.17 GHz / 2 Mbps / 1-10 km	Worldwide availability	High spectrum fee	HAN, NAN, AMI, DR
4G LTE-Advanced	800 MHz, 900 MHz, 1800 MHz, 2100 MHz, 2600 MHz, etc. / 300 Mbps / 1-10 km	Worldwide availability	High spectrum fee	HAN, NAN, AMI, DR, SCADA Backhaul
WiMAX	2.5 GHz, 3.5 GHz, 5.8 GHz / 75 Mbps / 1-5 km (Non-Line of Sight), 10-50 km (Line of Sight)	High data rate, long range	Not widespread	NAN, AMI, DR, SCADA Backhaul
WiFi	900 MHz, 2.4 GHz, 3.6 GHz, 5 GHz, 60 GHz / 600 Mbps / 30 m (indoor), 100-400 m (outdoor)	Worldwide standard, Easy deployment	Limited range	HAN, NAN, AMI
Wireless Mesh	900 MHz, 2.4 GHz, 5 GHz, 4.9 GHz / 300 Mbps / 10 km	High data rate	The waste of access spectrum for backhaul	NAN, AMI
Optical Fibre	180 THz - 330 THz / 100 Gbps / 100 km (9/125 single mode fibre at 1 Gbps)	Long range, High data rate	High deployment cost	WAN, WASA
Metro Ethernet	1.544 MHz, 2.048 MHz / 10 Gbps / 10 km –(in practice) 25 km	Affordable, Scalable, High data rate, Direct Ethernet	Relying on routing policies, Possibility of network outages due to loops	WAN, WASA
PLC	1-30 MHz / 3 Mbps / 1-3 km	Existing electrical wiring infrastructure is used.	Noisy and harsh channel environment	AMI
DSL	25 KHz - 1 MHz, 30 MHz / 100 Mbps / 5.5 km	Existing telephone wiring is used.	Potty connection if phone usage becomes heavy, Uploads are slower	AMI
Satellite / VSAT (Very Small Aperture Terminal)	4 GHz – 6 GHz (C-band), 12 GHz – 18 GHz (Ku-band), 18 GHz – 40 GHz (Ka-band) / 16 Mbps / Extensive coverage	Support for flexible data rates, Geographical availability, Location and time information based on the GPS	Expensive	SAS
ZigBee	868-915 MHz, 2.4 GHz / 250 Kbps / 30-50 m	Worldwide standard	Short range, low data rate	HAN, AMI
6LowPAN	868 MHz, 902-928 MHz, 2.4 GHz / 250 Kbps / 20 m	Native IPv6 support	Short range, low data rate	HAN, AMI
Z-Wave	908.42 MHz / 40 Kbps / 30 m	Ideal for smart home	Short range, low data rate	HAN, AMI
Bluetooth	2.4 GHz / 2 Mbps / 10-100 m	Supported by many types of devices	Short range	HAN

(Ferreira, Lampe, Newbury, & Swart, 2010; Gungor et al., 2013; Yan, Qian, Sharif, & Tipper, 2013; Baimel, Tapuchi, & Baimel, 2016)

Home Area Network (HAN)

Consumers value detailed information about their energy consumption costs, and are greatly interested in hardware/software-based tools that enable them to modify their energy consumption. Smart Grid technologies can address this need by empowering consumers to reduce their energy bill. Basically, a HAN is a dedicated network which enables smart meter connectivity with displays, load control devices and smart appliances, and this way provides electric utilities a platform to establish bidirectional communication with consumers' premises as shown in Figure 2. In sum, HANs can be described as a logical extension of smart meter deployment, dedicated to DSM and improvement in energy efficiency. Key components of a HAN are HEMS, In Home Display (IHD), hardware display device which can show meter-based consumption, bill-to-date information, and pricing signals, and smart appliances. Basically, smart appliances are any kinds of smart home appliances or appliances connected through smart plugs which can be controlled remotely either through IHD or web portals, and software systems such as AMI head-end, customer information system, self-service web portals, and mobile device management.

HANs provide many benefits. First, HANs allow Smart Grid infrastructure to benefit home-owners directly by assisting electric utilities in managing peak demand and enabling consumers to create schedules for switching of different devices based on criticality and predict their billing expenses (Gungor et al., 2011). Second, consumers can view consumption patterns of different devices installed in the premises in real-time, take decision regarding energy efficiency, monthly bill, demand response, and more, and switch to a different tariff plan. Third, consumers can view hourly energy consumption pattern of devices and design house-hold energy efficiency programs. Fourth, grid reliability can be provided through remotely controlled devices/ appliances. Flexibility in controlling demand through auto-schedules by electric utilities or manual control by consumers helps to eliminate potential blackouts. Finally, in the long run, net and peak energy consumption and consequently carbon footprint can be reduced.

Neighbourhood Area Network (NAN)

Neighbourhood Area Networks (NANs) allow Smart Grid infrastructure to connect smart meters and Distribution Automation (DA) equipment to WAN gateways. There are a number of specific uses for NAN communications. One of the main uses of NANs is to communicate with smart meters. Generally, electric utilities use meters from a single vendor in order to simplify communications from the utility to the meter. Nevertheless, by standardizing communications using open, global protocols, meters from different vendors can be interoperable and this way it is

Figure 2. Home Area Network

Neighbourhood Area Network   Home Area Network

possible to lower selection and deployment risk to electric utilities and ensure that all of these devices can work with data aggregation points and other devices (Kuzlu, Pipattanasomporn, & Rahman, 2014).

Wide Area Network (WAN)

A WAN aggregates data from multiple NANs and delivers it to the electric utility's private networks. Also, it enables long-haul communications among different data aggregation points of power generation plants, distributed energy resource stations, substations, T&D grids, and control centers (Gungor et al., 2011). In addition, it is responsible for providing the two-way network, needed for DA, substation communications and power quality monitoring. The WAN typically covers a very large area and can aggregate many supported devices. Hence, it requires hundreds of Mbps of data transmission.

RESEARCH CHALLENGES

With its distributed generators, a smart grid can exhibit complex operational regimes which make conventional management approaches no longer adequate (Yang, Barria, & Green, 2011). There are a number of research challenges that affect design, development and implementation of smart grid applications. First of all, having elaborately-defined communications and platform technology standards to enable smooth integration and interoperability between HANs, smart meters and other information technology systems is essential (Kuzlu, Pipattanasomporn, & Rahman, 2014). Second, in-home displays, smart appliances, smart meters and HEMS must be well-integrated with information technology systems found in utility such as meter data management, customer information system and AMI head-end, so that the true benefits of smart grid applications can be harvested. Third, there is a lack of well-defined regulation set regarding the transaction ownership of different device control operations including technological limitations in monitoring audit history and logging. Fourth, emerging smart grid technologies need time to return consistent and stable results in order to prove their reliability. Hence, during the stabilizing period, existing consumer awareness programs should be strengthened and focus on long term benefits to ensure the acceptance of those technologies. Specifically, consumers are not totally aware of the true potential of HAN, consumer awareness about HAN and change management is of key importance in increasing its acceptability. Fifth, in-house automatic device control mechanisms require availability of single line diagram of the electrical network and maintaining of device history within customers' premises. To sum up, both smart grid test environments and real world experiences of utilities show that the following aspects are critical to a result-oriented, purpose-built, successful smart grid implementation.

- Smart grid networks provide electric utilities with minimal administrative control over their operating environment. Since the environment is possibly going to change with time, and smart grid standards must enable the network to handle unpredictable interference, unexpected obstructions and other difficulties without requiring manual network adjustments (Tuna, Gungor, & Gulez, 2013).
- Compatibility with multiple communications technologies and interoperability between HEMS and different smart meter brands should be ensured for service-oriented, interoperable, scalable integration architecture. The integration architecture should natively support evolution of communications technologies and be backed by many partners.
- Different from the fact that although most communications networks today support communications between a human and a computer, smart grids use machine-to-machine (M2M) communications and support automated functions such as meter reading or outage response. Therefore, although smart grid networks must partly rely on peer-to-peer (P2P) communications, they must have the characteristics, such as redundancy and low latency, needed to accommodate grid automation.

- Major business processes should be independent of communications and strict operational standards should be maintained to boost consumer confidence.
- Vendors should be enabled to produce compatible products through standard certifications.
- An extensive audit control mechanism should be in place for participating systems.
- Devices must meet up-to-date information security standards to be able to address any kinds of data thefts, loss of privacy, and other security breaches (Rawat & Bajracharya, 2015).
- Using marketing campaigns across social media, a consumer awareness program must be conducted by utilities.

The increasing demand for wireless communications across all industries can positively impact the cost and performance of wireless utility networks. However, the increasing demand means increasing inferference and the increasing interference means higher latency and increased read failures on unlicensed communications networks such as Wi-Fi and RF mesh, although this is not an issue for licensed communications networks such as RF long-range radio, cellular and some WiMAX solutions at the expense of high costs and significant barriers (Baimel, Tapuchi, & Baimel, 2016). The impact of reliability concerns of a chosen solution must be considered especially for mission-critical smart grid applications (Bouhafs, Mackay, & Merabti, 2012). Since the choice of communications technology is very critical to provide reliable and efficient two-way communications and current communications infrastructures and standards are not capable of meeting the strict smart grid requirements such as range, rate, delay, and data security, cognitive radio enabled dynamic spectrum access is a candidate technology to reduce the communications expenses by reusing the existing wireless infrastructures and wireless spectrum to improve the overall performance (Kogias, Tuna, & Gungor, 2016). Dynamic spectrum access provides better throughput and faster delivery of information in smart grids leading to enhanced overall performance. Moreover, when dynamic spectrum access enabled communications is implemented in smart grids, devices can hop from a channel to the least interfered channel on the fly which results in jamming resistant communications.

Standards pose a significant challenge for electric utilities since the emergence of several interoperability standards has led to the widespread adoption of proprietary networking solutions and protocols. But integrating these proprietary networks is a significant challenge that will continue to exist with future versions of hardware and software. Related to this, electric utilities also have difficulty with choosing solutions, deciding where to deploy them, and picking vendors since they are concerned by the lack of communications expertise in-house (Asbery, Jiao, & Liao, 2016). Another challenges related to networking solutions are security of end-to-end networks and management of multiple communications networks prior to deploying new smart grid technologies.

FUTURE RESEARCH DIRECTIONS

Communications technology has been being used by electric utilities to support remote grid operations for a last couple of decades. But, with the smart grid transformation, they have already recognized that a single communications technology is not able to address all smart grid needs. For instance, while low-latency smart grid applications like DAS can be realized using broadband technologies, for targeted AMI deployments cellular technologies can be preferred. In this regard, to provide electric utilities a low-risk path for smart grid transformation the integration of different communications technologies into a common, device-independent platform is needed. Therefore, a unified, multi-technology, purpose-built communications platform should be developed to give electric utilities the simplicity of one platform with the flexibility of multiple technologies based on open standards and backed by extensive industry support (Baimel, Tapuchi, & Baimel, 2016). In a single deployment, such a platform can offer electric utilities the choice of mixing industry-leading communications technologies and help to integrate evolving technologies and landscapes of applications into a unified whole.

It is well-known that safety and reliability issues are top priorities for electric utilites. Since nowadays there are thousands of new potential points of failure due to new devices being added to the power grid, to ensure reliable power delivery, managing the increasingly complex communications network is critical (Gungor et al., 2013). This is more critical especially for the utilities faced with an expanding communications network. For such utilities, there is an urgent need for a comprehensive inventory management solution for its network assets. This solution can maintain accurate, up-to-date information on all network elements, attributes and configurations and thereby allows quick and efficient identification of network issues and impacted power grid elements.

Since most SCADA protocols were designed long before network security was considered as a relevant risk, and traditionally, grid devices and assets were mostly isolated or interconnected through private local networks, applying security measures on SCADA protocols was not considered as a priority. However, this is no longer the case and smart grids bring a new level of interconnection in which devices can be connected to large networks, wireless networks and the Internet (Tawde, Nivangune, & Sankhe, 2015). Hence, secure communications protocols must be used on SCADA devices in order to protect smart grids. Also, there is a need to develop policies to ensure periodic software and firmware updates to SCADA systems. Since these systems are now interconnected, and in many occasions passing through the Internet and other public networks, it is necessary to update them periodically to fix vulnerabilities and add new security measures to ensure the safety of smart grid communications from the origin and up to the destination.

To ensure that smart grid projects are successful, expertise should be provided for each phase of a smart grid process. After a smart grid is up and running, a versatile management suite can make the management of an entire smart grid infrastructure easier for electric utilities and keep it up and running for maximum benefits. However, such a suit should be built for flexibility and integration, highly scalable, and highly secure so that it can support any utility's vision.

CONCLUSION

In this chapter a detailed review of smart grid applications has been given and the use of various communications technologies for these applications has been discussed. Accordingly, communications standards employed in smart grid communications networks and existing and emerging communications solutions have been investigated. Finally in this chapter research challenges have been presented and open research issues have been outlined.

Since a smart grid is an automated electricity delivery and control system for an improved electricity supply chain running from a major power plant all the way inside our homes and offices, to enable desired sensing, monitoring, communication and control capabilities, smart grid communications infrastructure must address a number of key requirements. First, the communications infrastructure should match the technical requirements. Second, the communications infrastructure should not be subjected to external effects like unstable regulatory aspects, policy of access providers and consumer behaviour. Because, if the chosen solution is secured against such kind of effects, a stable environment needed to attract big investments can be expected. Third, the expenditures to attain the desired coverage, installation and operation should be as low as possible. Since day by day electric utilities deploy more equipment to support their expanding portfolios, the interoperability and standards problems will possibly worsen. Therefore, to maximise their investments, they need a network management solution to assist them with integrating proprietary networks, managing the health of numerous networks and supporting multiple applications and for monitoring and control purposes.

REFERENCES

Asbery, C. W., Jiao, X., & Liao, Y. (2016). Implementation guidance of smart grid communication. *Proceedings of 2016 North American Power Symposium (NAPS)*, 1-6.

Baimel, D., Tapuchi, S., & Baimel, N. (2016). Smart grid communication technologies-overview, research challenges and opportunities. *Proceedings of 2016 International Symposium on Power Electronics, Electrical Drives, Automation and Motion (SPEEDAM)*, 116-120. 10.1109/SPEEDAM.2016.7526014

Bouhafs, F., Mackay, M., & Merabti, M. (2012). Links to the future: Communication requirements and challenges in the smart grid. *IEEE Power & Energy Magazine*, *10*(1), 24–32. doi:10.1109/MPE.2011.943134

Bumiller, G., Lampe, L., & Hrasnica, H. (2010). Power line communications for large-scale control and automation systems. *IEEE Communications Magazine*, *48*(4), 106–113. doi:10.1109/MCOM.2010.5439083

Department of Energy. (2010). Communications Requirements of Smart Grid Technologies. Washington, DC: Academic Press.

Fan, Z., Kulkarni, P., Gormus, S., Efthymiou, C., Kalogridis, G., Sooriyabandara, M., ... Chin, W. H. (2013). Smart Grid Communications: Overview of Research Challenges, Solutions, and Standardization Activities. *IEEE Communications Surveys and Tutorials*, *15*(1), 21–38. doi:10.1109/SURV.2011.122211.00021

Ferreira, H., Lampe, L., Newbury, J., & Swart, T. (2010). *Power Line Communications*. New York: Wiley. doi:10.1002/9780470661291

Garcia-Hernandez, J. (2011). An analysis of communications and networking technologies for the smart grid. *Proceedings of CIGRE International Symposium, The Electric Power System of the Future*.

Gezer, C., & Buratti, C. (2011). A ZigBee smart energy implementation for energy efficient buildings. *Proceedings of 2011 IEEE 73rd Vehicular Technology Conference (VTC Spring)*, 1-5. 10.1109/VETECS.2011.5956726

Giannuzzi, G. M., Pisani, C., Vaccaro, A., & Villacci, D. (2015). Overhead transmission lines dynamic line rating estimation in WAMS environments. *Proceedings of 2015 International Conference on Clean Electrical Power (ICCEP)*, 165-169. 10.1109/ICCEP.2015.7177618

Gomes, D., Colunga, R., Gupta, P., & Balasubramanian, A. (2015). Distribution automation case study: Rapid fault detection, isolation, and power restoration for a reliable underground distribution system. *Proceedings of 2015 68th Annual Conference for Protective Relay Engineers*. 10.1109/CPRE.2015.7102176

Gungor, V. C., Sahin, D., Kocak, T., Ergut, S., Buccella, C., Cecati, C., & Hancke, G. P. (2011). Smart grid technologies: Communication technologies and standards. *IEEE Transactions on Industrial Informatics, 7*(4), 529–539. doi:10.1109/TII.2011.2166794

Gungor, V. C., Sahin, D., Kocak, T., Ergut, S., Buccella, C., Cecati, C., & Hancke, G. P. (2013). A survey on smart grid potential applications and communication requirements. *IEEE Transactions on Industrial Informatics, 9*(1), 28–42. doi:10.1109/TII.2012.2218253

Hart, D. G. (2008). Using AMI to realize the Smart Grid. *Proceedings of 2008 IEEE Power and Energy Society General Meeting - Conversion and Delivery of Electrical Energy in the 21st Century*. 10.1109/PES.2008.4596961

Kogias, D., Tuna, G., & Gungor, V. C. (2016). Cognitive Radio. In H. T. Mouftah & M. Erol-Kantarci (Eds.), *Smart Grid: Networking, Data Management, and Business Models*. Boca Raton, FL: CRC Press.

Kuzlu, M., Pipattanasomporn, M., & Rahman, S. (2014). Communication network requirements for major smart grid applications in HAN, NAN and WAN. *Computer Networks, 67*, 74–88. doi:10.1016/j.comnet.2014.03.029

Li, Z., Wang, P., Chu, Z., Zhu, H., Sun, Z., & Li, Y. (2013). A three-phase 10 kVAC-750 VDC power electronic transformer for smart distribution grid. *Proceedings of 2013 15th European Conference on Power Electronics and Applications (EPE)*. 10.1109/EPE.2013.6631810

Liserre, M., Sauter, T., & Hung, J. Y. (2010). Future energy systems: Integrating renewable energy sources into the smart power grid through industrial electronics. *IEEE Industrial Electronics Magazine, 4*(1), 18–37. doi:10.1109/MIE.2010.935861

Lu, D., Kanchev, H., Colas, F., Lazarov, V., & Francois, B. (2011). Energy management and operational planning of a microgrid with a PV-based active generator for smart grid applications. *IEEE Transactions on Industrial Electronics, 58*(10), 4583–4592. doi:10.1109/TIE.2011.2119451

Mahmood, A., Aamir, M., & Anis, M. I. (2008). Design and implementation of AMR Smart Grid System. *Proceedings of 2008 IEEE Canada Electric Power Conference (EPEC 2008)*. 10.1109/EPC.2008.4763340

Matsumoto, S., Serizawa, Y., Fujikawa, F., Shioyama, T., Ishihara, Y., Katayama, S., ... Ishibashi, A. (2012). Wide-Area Situational Awareness (WASA) system based upon international standards. *Proceedings of 11th International Conference on Developments in Power Systems Protection (DPSP 2012)*. 10.1049/cp.2012.0032

Meiling, S., Schmidt, T. C., & Steinbach, T. (2015). On performance and robustness of internet-based smart grid communication: A case study for Germany. *Proceedings of 2015 IEEE International Conference on Smart Grid Communications (SmartGridComm)*. 10.1109/SmartGridComm.2015.7436316

Meliopoulos, A. P. S., Polymeneas, E., Tan, Z., Huang, R., & Zhao, D. (2013). Advanced Distribution Management System. *IEEE Transactions on Smart Grid*, *4*(4), 2109–2117. doi:10.1109/TSG.2013.2261564

Palensky, P., & Dietrich, D. (2011). Demand side management: Demand response, intelligent energy systems, and smart loads. *IEEE Transactions on Industrial Informatics*, *7*(3), 381–388. doi:10.1109/TII.2011.2158841

Rawat, D. B., & Bajracharya, C. (2015). Cyber security for smart grid systems: Status, challenges and perspectives. *Proceedings of SoutheastCon 2015*, 1-6. 10.1109/SECON.2015.7132891

Saber, A. Y., & Venayagamoorthy, G. K. (2011). Plug-in vehicles and renewable energy sources for cost and emission reductions. *IEEE Transactions on Industrial Electronics*, *58*(4), 1229–1238. doi:10.1109/TIE.2010.2047828

Sauter, T., & Lobashov, M. (2011). End-to-End communication architecture for smart grids. *IEEE Transactions on Industrial Electronics*, *58*(4), 1218–1228. doi:10.1109/TIE.2010.2070771

Su, W., Eichi, H., Zeng, W., & Chow, M. Y. (2012). A survey on the electrification of transportation in a smart grid environment. *IEEE Transactions on Industrial Informatics*, *8*(1), 1–10. doi:10.1109/TII.2011.2172454

Tawde, R., Nivangune, A., & Sankhe, M. (2015). Cyber security in smart grid SCADA automation systems. *Proceedings of 2015 International Conference on Innovations in Information, Embedded and Communication Systems (ICIIECS)*, 1-5. 10.1109/ICIIECS.2015.7192918

Tuna, G., Gungor, V. C., & Gulez, K. (2013). Wireless sensor networks for smart grid applications: A case study on link reliability and node lifetime evaluations in power distribution systems. *International Journal of Distributed Sensor Networks*, *2013*, 1–11.

Wenpeng, L., Sharp, D., & Lancashire, S. (2010). Smart grid communication network capacity planning for power utilities. *Proceedings of 2010 IEEE PES Transmission and Distribution Conference and Exposition*, 1-4. 10.1109/TDC.2010.5484223

Yan, Y., Qian, Y., Sharif, H., & Tipper, D. (2013). A Survey on Smart Grid Communication Infrastructures: Motivations, Requirements and Challenges. *IEEE Communications Surveys and Tutorials*, *15*(1), 5–20. doi:10.1109/SURV.2012.021312.00034

Yang, Q., Barria, J. A., & Green, T. C. (2011). Communication infrastructures for distributed control of power distribution networks. *IEEE Transactions on Industrial Informatics*, *7*(2), 316–327. doi:10.1109/TII.2011.2123903

Yigit, M., Gungor, V. C., Tuna, G., Rangoussi, M., & Fadel, E. (2014). Power line communication technologies for smart grid applications: A review of advances and challenges. *Computer Networks*, *70*, 366–383. doi:10.1016/j.comnet.2014.06.005

Zaballos, A., Vallejo, A., & Selga, J. M. (2011). Heterogeneous communication architecture for the smart grid. *IEEE Network*, *25*(5), 30–37. doi:10.1109/MNET.2011.6033033

KEY TERMS AND DEFINITIONS

Advanced Metering Infrastructure (AMI): It is a fully integrated system of smart meters, communication networks, and data management systems which enables bidirectional communication between electric utilities and customers.

Automatic Meter Reading (AMR): A technology that automatically collects status, diagnostic and consumption data from energy metering devices. This data is transferred to the utility server for billing purposes; to analyze usage and manage consumption, and to identify or resolve technical problems.

Demand Response (DR): It refers to programs which aggregate small reductions from many different customers and then treat that reduction as a single block of power.

Distributed Energy Resources (DER): They are small-scale power generation sources located near to where electricity is used. They provide an alternative to or an enhancement of the traditional power grid.

Distribution Automation System (DAS): A technology that can be implemented on the power grid's distribution system of local power lines and neighbourhood substations. It enhances the reliability of a power grid with real-time monitoring and intelligent control capabilities.

Home Area Network (HAN): A network deployed and operated within a home or small office to enable the communication and sharing of resources between computers, tablets, phones, and other devices over a network connection. It can employ several connectivity technologies, wired and/or wireless, related to the specific subsystems installed in the home. Typical HAN subsystems include computer networks, Heating, Ventilating and Air Conditioning (HVAC) systems, alarm systems, and other types of subsystems related to the smart grid applications, such as local energy generation systems, energy management systems and V2G systems.

Home Energy Management System (HEMS): A system that effectively utilizes renewable energy by visualizing load equipment information in the home and properly controlling it. According to the information of supply and demand arrangement requests from community energy management system, it contributes to the peak shifting and load shifting processes.

Neighbourhood Area Network (NAN): It allows smart grid infrastructure to connect smart meters and DA equipment to WAN gateways. There are many specific uses for NAN communication.

Power Line Communication (PLC): A communication technology that enables sending data over existing power cables.

Renewable Enery: Energy generated from natural processes that are continuously replenished such as wind, sunlight, tides, geothermal heat, and various forms of biomass. This energy cannot be exhausted and is constantly renewed.

Smart Grid: A modern power grid infrastructure for enhanced efficiency, improved reliability and high safety, with seamless integration of renewable and alternative energy sources, through advanced control strategies and modern communication technologies.

Smart Meter: A new kind of meter which can digitally send meter readings to the electricity supplier to ensure more accurate electricity bills.

Substation Automation System (SAS): A system that performs data acquisition, remote communication, supervision control, protection and fault evaluation.

Vehicle to Grid System (V2G): A system that enables the flow of power between an electric system or grid and electric drive vehicles.

Wide Area Network (WAN): It provides network connectivity to the utility's control center over either a service provider network or over a utility-owned network traversing the primary substations in many cases.

Wide Area Situational Awareness (WASA): A set of technologies designed and implemented to improve the monitoring of a power grid across large geographic areas. It efficiently provides grid operators with a dynamic and broad picture of the functioning of the power grid.

Wireless Mesh: Wireless mesh networks are composed of cooperating radio nodes organized in a mesh topology. The link communication technology from one hop to another can be standardized (e.g., IEEE 802.11 series or IEEE 802.15.4) or proprietary (e.g., OFDM, FHSS technologies). Thanks to their mesh properties along with self-setup and self-healing mechanisms, they inherently offer operation easiness and required redundancy for fixed applications.

Chapter 9

Understanding Smart City Solutions in Turkish Cities From the Perspective of Sustainability

H. Filiz Alkan Meshur
Selcuk University, Turkey

ABSTRACT

The purpose of this chapter is to analyze the concept of smart city and its potential solutions to correct urban problems. Smart city practices and solutions have been investigated through the lens of a sustainable perspective. As the general practices in the global scale were examined, particular focus has been directed to smart city practices in Turkey and applicable suggestions have been developed. A number of cities in Turkey rank the lowest in the list of livable cities index. Consequential to the rapidly rising population ratios, the quality of provided services declines; economic and social life in cities are adversely affected and brand images of cities are deteriorated. With the implementation of smart city practices, such problems could be corrected, and these cities could gain competitive advantage over their rivals. The key component of this smart administration is to most effectively utilize information and communication technologies during each single step of this process.

DOI: 10.4018/978-1-5225-3996-4.ch009

Copyright © 2018, IGI Global. Copying or distributing in print or electronic forms without written permission of IGI Global is prohibited.

CONCEPTUAL FRAMEWORK

According to recent projections, the global population in year 2050 will soar to 10 billion of people half of which will be populated in cities. It is nevertheless a potential spot of concern that such rapid increase in global population ratio would bring with itself a range of problematic issues. As widely agreed, depletion of limited resources would inevitably lead to scarcity of resources for the prospective generations. Although population ratio is on the decrease in developed states it is continually on the rise in underdeveloped states which correlates to the acceleration in the population rates of deprived classes. In parallel with the enhanced pressure towards cities the economic, social and sub-structural problems have been increasingly multiplied (Figure 1).

According to the United Nations' World Urbanization Prospects report of year 2015, six out of every ten people are expected to live in urban areas by 2030 and this rate will increase to 66% in 2050 (Figure 2).

Smart City Concept and Features

Urbanization leads to narrowing of the spaces opened for both the expansion of urban areas and for other uses since the requirement of finding new areas for the cities and those who come to settle in the city. Moreover, this considerable energy consumption and carbon monoxide gives rise to gases such as greenhouse gases to affect the environment. For this; both ecological and technological to the cities (smart cities) are needed (Ayber, 2016).

Figure 1. World Population (UN, 2015)

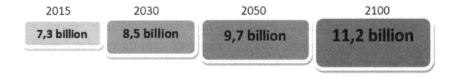

Figure 2. The ratio of people living in cities (UN, 2016)

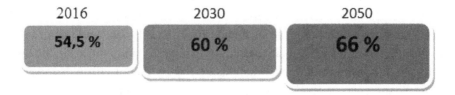

Smarter City uses digital technology and information and communication technologies (ICT) to better quality and performance this engage more effectively and actively with its citizen. The Smart City includes government services, transport, traffic management, energy, health care, water and waste. The smarter city applications are major goal of improving the management and transforming the urban areas. The major technological, economic and environmental changes have generated interest in smart cities. A city can be defined as 'smart' when investments in human and social capital and traditional (transport) and modern (ICT) communication infrastructure fuel sustainable economic development and a high quality of life, with a wise management of natural resources, through participatory action and engagement (Kumar, 2015).

A smart city uses information and communications technology (ICT) to enhance its livability, workability and sustainability. In simplest terms, there are three parts to that job: collecting, communicating and crunching. First, a smart city collects information about itself through sensors, other devices and existing systems. Next, it communicates that data using wired or wireless networks. Third, it crunches (analyzes) that data to understand what's happening now and what's likely to happen next (Smart Cities Readiness Guide, 2015).

While smart city vision is being explained as the integration of communication and automation systems (outdoor sensors, mobile terminals, etc.) with special and public spaces, on the other hand, it reflects the target of making the historical city eco-friendly and energy efficient by integrating smart technologies into city. With ensuring green energy substructure (solar panels generating heat and electricity, home wind turbines, electric vehicle charging stations, efficient energy distribution and control with smart grid, etc.), it is aimed to reduce carbon footprint of the city. According to Lugaric et al. (2010) the components of a smart city are reliable energy and water supply, efficient public administration, 7/24 access to public data, high level of intellectual social capital, competitor manufacturer and open local economy (Sınmaz, 2013). With smart city applications, it can be found solutions to basic problems of the cities. The basic problems faced by cities can be seen in Table 1 (Akgül 2013).

In this sense, in the ideal of smart/efficient city, energy efficiency and technological harmony that comes forefront in recent years and an efficient city form must be integrated (compact and self-sufficient). Besides, local economy and social benefit concepts are power factors for the sustainability of an efficient settlement. In this framework, smart cities are the cities used for using efficiently the available urban area in face of consumption factors increasing day by day in all fields today and adopting the strategies of developing preventions for energy consumption, increasing

Table 1. Basic problems faced by the cities

Area	Basic Problems
Transportation	The current transportation substructure remains insufficient for the number of vehicles that is increasing with the population and the time elapses in the traffic increase. The increased time leads to job losses, reduction in the efficiency increasing in the transportation costs and increasing in harmful exhaust gas emissions. The number of traffic accidents increases; as a result health expenses and costs increase as well as loss of life.
Energy	With the increasing energy demand, more expensive and inefficient energy resources are used. Losses in the distribution and illegal use of electricity use increase.
Water	Water resources around the cities could not meet the needs of the cities and losses the feature of renewable. The pollution caused by the cities threats clean water resources. Due to population intensity, the control of epidemic illnesses becomes difficult. The environmental effects of urban life negatively affect public health.
Health	Due to the problems especially in the transportation, there may be some delays in emergency interventions on-site and in time. Due to population intensity, the control of epidemic illnesses becomes difficult.
Services	With the increase in the number of people living there, giving services for local and central administrations has become difficult and the quality of the services and therefore life quality in the cities has decreased.
Environment	The cities consume renewable resources rapidly and the environmental problems such as increase in the number of vehicles, air and water pollution generate big threats for the people living in the cities. Unorganized and unplanned urbanization bring the problems such as infrastructure and collecting and storing solid wastes along with their own problems.
Security	Population density and income difference in the cities leads to increase crime ratio. It cannot generate preventive solutions for security problems in time and on-site.

the life quality and supporting local economic potentials and using the developing technology to design and implement these strategies (Sınmaz, 2013).

The performances of the cities depends on collection of the information, the quality, processing and conveyance of the information (human and social capital), instead of the substructure they own (physical capital). Today, the protection of brand equity and competitiveness of the cities needs human resource and social capital. In this context, the concept of smart city drawing the framework of the service components of modern cities and the importance of efficient use of ICTs and resources has begun to be used increasingly in the strategies and service presentations of the cities for last twenty years (Smart Cities in Europe, 2016).

In addition to the concept of smart city, 'smart city solutions/applications' are ICT supported applications basically integrated into the information technologies substructure of the cities such as City Information Systems (CIS) and Geographical Information Systems (GIS) and that facilitates the presentation and monitoring many services such as energy, water, transportation, health, education, security to

Table 2. Solution approaches to urban problems with smart systems

Solution Approaches to Urban Problems With Smart Systems
Natural Resources and Energy   *Smart Grids*   Electric networks that add account to the behavior of interconnected users for sustainable, economical and secure electricity supply (Intelligent networks must be self-monitoring and resistant to system irregularities).   *Public Lighting*   Street lamps used to illuminate public spaces also perform other functions such as air pollution control or wireless internet connection. (Operating costs can be reduced through centralized management systems communicating with street lights, instantaneous information about the weather can be provided).   *'Green' Energy*   Utilizing renewable natural sources such as the sun, water and wind.   *Waste Management*   Development of waste collection and recycling methods that will minimize the negative effects on the society and the environment.   *Water Management*   Analysis and management of water quantity and quality, especially in agricultural and industrial use, with hydrologic cycle stages.   *Food and Agriculture*   Establishment of wireless receiver networks to monitor the conditions of agricultural crop areas and manage cultivation processes (By combining moisture, temperature and light receivers within integrated systems, the risks of frost or plant disease can be reduced and irrigation needs can be optimally regulated.
Transportation and Mobility   *City Logistics*   Improving the logistics flow in the city by effectively integrating traffic situation, geographical and environmental factors and working life requirements.   *Mobility Information*   Dynamic and multi-mode information systems are used to increase traffic and transport efficiency and the quality of the transportation experience. Information is provided concurrently prior to travel and during travel. These systems take human behavior into account.
Buildings   *Building Services and Management*   The control of electrical and mechanical equipments of buildings with central information systems and integration of systems such as cleaning, maintenance and security between buildings and buildings.   *Home quality*   The harmonization of houses with sustainable energy and environmental management of the city.

(Akıllı Kent Masabaşı Araştırması. Kamu, 2016b)

enable making decision based on real-time information and increasing its quality and increasing its quality.

Smart cities ensuring sustainable economic development and high level of life quality have reached to an advance level in many fields such as environment, human resource, education, public services and transportation. The basic elements of catching the development in these areas are in need of qualified human resource, social capital and efficient use of ICT. As of its scope, basic features of the concept of 'smart city' can be summarized as follows (T.C. Kalkınma Bakanlığı., 2013).

- Using the sub-structures that are in contact with each other for increasing economic and political efficiency and ensuring social, cultural and urban development,
- Increasing socio-economic development of the cities by creating business-oriented approach (creating new business opportunities in the cities and developing them) and through them,
- Ensuring equal use of the social content of city residents in different sections and regions and of public services,
- In information oriented and globalizing economy, being drawn of the qualified human resource to the cities and forming creative culture environment,
- In urban development, the importance of social capital and being learnt the new technologies by the urban residents, adopting them and contributing to generating new solutions by the citizens for cities through the structures such as live laboratories,
- Ensuring social and environmental sustainability and using resources efficiently.

Figure 3 presents smart city components; smart economy, smart people, smart governance, smart mobility, smart environment and smart living. "Smart economy refers to the ability of the economy to be flexible to the labor market conditions and capable of transforming itself into productive units whatever the conditions maybe. Smart people refer to people in the city who have adequate level of education and skill so that they can contribute to the development and progress of the country. Smart Governance refers to a transparent government which allows participation of public in decision making. Smart mobility refers to safe and sustainable transportation system that is accessible to all members of the society and use of ICT infrastructure for reduction in travel time. Smart environment refers to the sustainable resource management and environmental protection. Smart living refers to good quality of life for all individuals who have easy access to health and education facilities and proper physical infrastructure. It denotes a society which is socially cohesive. All these components make a city more inter connected, efficient and intelligent" (Smart Cities Mission, 2016).

In addition, creating convenient life environments by using ICTs for disabled people must be an important target. Smart city solutions and transportation systems have great importance in increasing life quality of disabled people. With smart city systems, the easy movement of the elderly and disabled people will be ensured in many fields from transportation to social and cultural activities.

Smart city components and elements are indicated in Table 3.

The goal of UNECE (United Nations Economic Commission for Europe) for a smart city for all can be seen in Table 4.

Figure 3. Components of smart cities

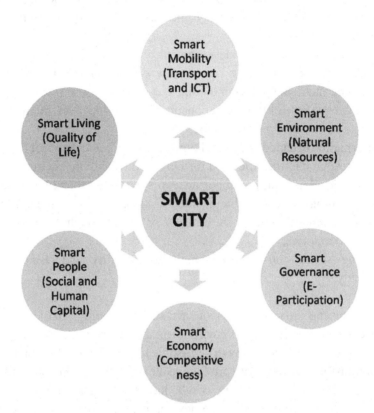

In the case that the urbanization is uncontrolled, the probability to encounter with some problems such as insufficient sub-structure, illegal housing and various demographic structuring. Such kind of developments in the world reveals the necessity of smart cities. This case has led to the formation of European smart cities and societies in Europe (European Innovation Partnership-EIP). The most important duty of EIP is to apply innovative and technological solutions in the cities for the formation of sustainable cities. This approach naturally reveals the idea of sustainable control. In this scope, the foundations such as housing and land administration unit of United Nations Economy Commission, UNECE, EU HABITAT offices, FIG, Urban Planners Organization, Dubai Real Estate Institute, Europe Environment Agency, OECD, and EBC decided to arrange a common project. With the concept of smart cities in the project, it has been aimed to implement sustainable urban development principles and applications including new business models to increase capital, technology and administration skills in order to exceed application difficulties having low middle income economy. For these struggles, the committee allowed for up to 2020 (URL, 1).

Table 3. Smart city components and elements

Smart Economy (Competitiveness)	Smart Person (Human Capital)	Smart Life (Quality of Life)	Smart Governance (Participation)	Smart Mobility (Transportation)	Smart Environment (Natural Resources)
• Innovative spirit • Entrepreneurship • Economic appearance and brands • Efficiency • Flexibility of the labor market • International relations • Public and private sector cooperation • Competitiveness of the city	• Qualification level • Capacity standard • Lifelong learning level • Social and ethnic diversity • Flexibility • Creativity • Open-mindedness • Participation • Having professional skills • Intelligent management	• Social facilities • Health care • Health situation • Personal security • Conditions of shelter • Training facilities • Touristic attraction • Social cohesion • Smart / green buildings	• Social participation in decision making • Transparent management • Digital infrastructure • Public services and social services • Emergency systems • Political strategy and perspectives • Politics and legislation • Urban digitization • Urban data tracking and measurement	• Travel preferences • Urban access • Intercity access • Information communication technology infrastructure • Innovative and safe transport systems • Traffic density, energy efficiency and carbon emissions	• Attractiveness of the natural conditions • Air pollution • Environmental Protection • Sustainable resource management • Energy efficiency • Renewable energy • Decrease in waste rate

(Akıllı şehir yol haritası Vodafone., 2016a)

Table 4. The goal of UNECE (United Nations Economic Commission for Europe) for a smart city for all

Make Cities and Human Settlements Inclusive, Safe, Resilient and Sustainable	
Main Challenges	**Smart City Should**
• High energy consumption • High quantities GHG emissions • Overpopulation and critical living conditions for citizens • Informal settlements • Conversion of land and green spaces into constructions and buildings • Mobility problems and lack of infrastructures and services • Housing issues	• ensure access to adequate and affordable housing • provide access to safe, affordable and sustainable transport systems • enhance inclusive and sustainable urbanization • safeguard the world's cultural and natural heritage • reduce the number of deaths, displacement and losses caused by disasters • reduce their environmental impact • provide universal access to safe and accessible green and public spaces

(Carrioera, 2015)

Sustainable Smart Cities

The sustainability concept is seen to start gaining importance with Brundtland Report signed in 1987. According to this report, sustainability means 'meeting the needs of today by taking into account the needs of next generations in a healthy and balanced way' and gains a contradictory and ambiguous meaning in ecological sense by orienting to basically development targets. Today, sustainable development is accepted to be a long-termed concept having-actually contradictory with each other most of the time-economic, social, political and environmental dimensions and the requirement to establish the discussion on the mutual interactions of these dimensions are mentioned (UN, 1987).

Using ICTs in the information and communication technology sector as well as other sectors, the provision of energy efficiency and the reduction of carbon emission constitute the basis of Green Information Technology. States have adopted information and communication technologies and environmental policies in different fields. In these policies, R&D activities, innovation, green knowledge, the application of information and communication technologies, and the creation of competence and awareness are key areas of policy (T.C. Kalkınma Bakanlığı., 2013).

The concerns about high energy prices and climate change have been the driving forces behind the interest in energy efficiency. The world has faced an oil crisis in the 1970s and for this reason the manufacturing sector has focused on energy efficiency in those years.

This interest has come to the present day again due to three main reasons. The first is the dramatic increase in energy prices and the significant energy savings that will be provided. Second, there are widespread interruptions due to increased supply of energy demand in some regions. This makes it capable of critical maintenance for energy-saving businesses. The third reason is the growing pressure to reduce energy consumption on companies with the highest CO_2 emissions, leading to fueling societal concerns of CO_2 emissions. The enactment of legislation that encompasses CO_2 emissions at various locations is an important incentive for companies to reduce their energy consumption. With energy efficiency initiatives, companies are making both right decisions and saving costs (McKinsey and Company, 2010).

In this context, the smart manufacturing approach is becoming increasingly important for companies seeking to improve production processes and improve energy efficiency. The Smart Production Approach, which can be defined as improvement of production processes through reduction of waste and energy consumption by utilizing technology, supports the following activities (T.C. Kalkınma Bakanlığı., 2013);

- Monitoring and management of production process,
- Both early design and desorption of products during the distribution phase,
- Logistics improvements in end-product delivery for end-users and consumers.

Sustainable approaches to the smart city are indicated in Table 5.

Innovative and intelligent solutions are needed to meet the growing expectations in a sustainable way. The concept of smart cities is a journey that is shaped by each city's own needs. Besides the mega-cities, efforts to raise the quality of service and quality of life in accordance with the needs of less populated, immigrant, historical or newly formed cities are also included in the concept of smart urbanism. The expectations and needs of new generations shaped in the light of technological developments should be considered in the world of tomorrow. For example, the technologies we can not even imagine in our childhood may have become a basic necessity for today's youth. Therefore, the concept of being ready for tomorrow requires a holistic approach, effort and investment (*Akıllı şehir yol haritası Vodafone.*, 2016a).

SMART CITY SOLUTIONS IN THE WORLD

According to OECD figures, it is expected to invest 70 trillion dollars for the infrastructure of smart cities in the world. Accordingly, making investment between 4-6 trillion dollars on average for each year in smart cities is in question. According to the estimations, there will be 36 mega cities and 26 smart cities up to 2025. From this point of view, being smart city will bring not only saving but also image and reputation. The way of leading in city marketing and regional development passes through this image and reputation (Yilmaz, 2015).

Table 5. Sustainable approaches to the smart city

The Reduction of Energy Consumption	- The reduction of motor vehicle usage - The creation of pedestrian and bicycle routes - Low energy use in lighting (solar energy, LED lighting, etc.)
The Reduction of Resource Consumption	- The importance of recycling materials to be selected in implementation projects - The use of permeable paving stones
The Effective Use of Natural Resources	- Rain water should be used in various places by providing recycling of them
Supporting Livability	- Collecting recycling waste - Supporting street art - The reduction of CO_2 emissions

When the country's policies and regional policies are examined, the studies on smart cities are seen to gain importance. European Union sustaining studies effectively in the field of smart cities aims to solve the problems of the cities in Europe with innovative IT applications with the initiative of 'The European Innovation Partnership on Smart Cities and Communities' (EIP-SCC) they formed recently. On EU smart city applications, EU provides financial support to many studies on smart city applications especially in the field of energy efficiency. In addition to this, 'live laboratory' in which the citizens are drawn into innovative process is commonly used in Europe for smart city applications.

Smart cities are one of the basic strategy areas of 12th five-year economic development plan of Chinese government (2011-2015) and high amount of public finance has been allocated on this matter and arrangements have been made. For the realization of smart city solutions, it is foreseen to allocate resource as much as 12,7 billion dollars between the years of 2013 and 2015.

British Technological Strategy Commission aiming to increase growth and efficiency through technological innovative solutions initiated a competition in the value of 25 million pounds for the realization of smart city solutions integrated with big-scaled and integrated with the other information technologies substructure systems of the city in 2012. The knowledge and implementation solutions to be gained as a result of the project have been planned to be opened for use of the other cities in England (T.C. Kalkınma Bakanlığı., 2013). London draws attention to its sustainable investments in becoming a smart city. For example, the traffic congestion tax that is applied to avoid the traffic congestion in the city is an interesting solution. An agreement has also been signed with the operator O2 for the establishment of the largest wireless network in Europe in the city. According to 2012 data in the list of the smartest cities of the world, only Istanbul takes place from Turkey. In the list on which Istanbul is in the 68th row, the title of the smartest city of the world belongs to Vienna. Looking to the Smart City Applications of the World, it is seen that the USA and EU countries are leading (Yilmaz, 2015). Vienna is one of the leading cities in the process of being a smart city with top policy workings such as Smart City Vision 2050, Road Map 2020 and Action Plan 2012-2015. The city administration is following these initiatives. In addition, the Viennese architects act as stakeholders and cooperate with the municipalities on issues such as reducing carbon consumption, facilitating city planning and transportation. Also, Copenhagen is one of the smartest cities in Europe. In order to achieve this, sustainability investments are emphasized. For example, efforts to reduce carbon footprint to zero can be made in the city by 2025. For this purpose, the city population is currently using bicycles at 40 percent for transportation (Digital Age, 2014).

Looking to the Smart City Implementations in the world, it is seen that some implementations take the eco-technology on the basis. For example; a new eco-tech settlement area implementation was built in the Milton city of Canada. The thing desired to be made in this city with approximately 45.000 populations is a self-supporting eco-city design with 4000 population. In New Zealand, 'Waitakere' city has entered into a formation of multi-participants since 1997. An action plan and network named as 'Waitakere Eco-Tech Action-WETA' and including the Municipality, research institutes and civil society organizations have been found. 'WETA supports sustainable future' is the slogan of this city. On the basis of this action plan, there is being used, understood and made use of IT by all of those who are settled in Waitakere city (Akgül, 2013).

Amsterdam city information and communication technologies in Holland are realized the projects about ICT based life and working, e-Participation, open data and mobile applications, energy efficiency, smart buildings, smart counters, remote health services and smart network subject through the institutions more than 70 formed by the partnership of private sector, local administration and research foundations. European Commission Regional Development Fund, private sector and public sector transfer resource to the projects and 'live laboratories' application which draws those who live in the cities into the innovation process is used in the realization of smart city applications in Amsterdam (T.C. Kalkınma Bakanlığı., 2013; URL, 6). The most innovative project that draws attention to the fact that the city is a smart city is the project related to Amsterdam ArenA Stadium. Through the signed cooperation agreement, it will be possible to manage the spectators in the stadium in an appropriate manner with intelligent ICT solutions. For the stadium, network management and connectivity enhancements will also be made via smartphones and tablets (Digital Age, 2014).

It has been seen that Barcelona has cooperated with private sector stakeholders to implement the Smart Cities Plan. In this respect, cooperation has been made with the telecom operator and the technology company for the whole plan (*Akıllı şehir yol haritası Vodafone.*, 2016a). The city government is demonstrating its adaptability to the smart city concept with its investments in low carbon consumption. To this end, the world's first solar thermal energy consumption instruction manual has been issued. Parallel to this, the project named LIVE EV Project was launched and the charging infrastructure was prepared. Also, a laboratory agreement for a widespread smart city innovation was signed in Barcelona (Digital Age, 2014).

Songdo City in South Korea is another example where smart city applications are realized. Songdo City is a new city recreated in a 1.500 decares of area. In the city whose construction has been planned to be finished in 2015, there will be 80.000 houses. In the concept of sustainable city, 40% of the areas in the city constructed are expected to be formed by green areas. Within the scope of the project, fiber wide band access will be provided throughout the city, video conference system screens (tele-presence) will be available in the houses, offices and public buildings and sensors will be used intensively in the applications such as lightening, transportation, energy, water management, security, smart buildings. It is realized with the partnership of Songdo City local administration and private sector partnership and with mainly the financing of private sector and local administration (T.C. Kalkınma Bakanlığı., 2013; Songdo IBD).

Tokyo as one of the most important smart cities in Asia, is proposing smart solutions for suburban areas. The city government has cooperated with Panasonic, Accenture and Tokyo Gas to develop intelligent applications for houses, solar energy panels, storage units and electric power networks. Tokyo is also attracting attention with smart mobile applications it has developed for residents (Digital Age, 2014).

Projects about water and electricity counters, smart transportation and traffic management and emergency case intervention are carried out Shenyang city in China with the participation of private sector, local administration and research institutions and mostly the financing of local administration (T.C. Kalkınma Bakanlığı., 2013).

"The project at Meixi Lake is an example of a start-up "smart" city planned for China. The same real-estate developer who took on the Songdo project for the Korean Government entered into an agreement with the Changsha Municipal People's Government of Hunan Province to plan, develop and operate Meixi Lake District, a state-of-the-art, new ecological city located on 1,675 acres in Changsha, the capital city of Hunan Province" (Belissent 2010).

Toronto recognized as North America's smartest city, stands out for its incentive to use low carbon. In Toronto, not only the public sector but also the private sector are working intensively for this purpose. Current developments can be given to the operation of garbage trucks with natural gas in the city. For this, the garbage collected from the garbage collection area of the city is used.

SMART CITY SOLUTIONS IN TURKEY

According to the research company Frost & Sullivan, it is estimated that by 2025 there will be more than 26 smart cities in Turkey. The cities that have invested in a few, even if not in the majority of the areas considered as smart city criteria, are called 'sustainable cities'. It is predicted that eight cities (Istanbul, Bursa, Ankara, Eskişehir, Izmir, Denizli, Antalya and Adana) will be in sustainable city level in Turkey (Belissent, 2010).

Expert staff number qualified about smart city in Turkey is insufficient. In this sense, narrow-scoped applications are made. The first implementations are the project of foundation of eco-tech settlement place called as Information Valley

Project initiated in Yalova in 2000. Information Valley Projects were brought to the agenda also by Bursa, Kocaeli, Ankara and the other cities (Akgül, 2013).

The first smart city application supported by the EU will start in Eskişehir. 'Eskişehir Tepebaşı Municipality Life Village Project' will be an important experience. The project seeks to provide sustainability through improvements in transport, communications and information technology. Within the scope of the project, with the help of technology, energy efficiency has been increased and efficient use of resources has been targeted. In this respect, Tepebaşı aims to reduce greenhouse gas emission rates by 80 percent. In Turkey, after 2009, the EU has put forward a considerable amount of budgeted projects from the Information and Communication Technologies Policy Support Program (ICT PSP), one of the three programs under the Competitiveness and Innovation Framework Program (CIP). In particular, projects launched in 2012 and the ICT PSP program will include 'Information and Communication Technologies (ICT) for Low Carbon Economy and Intelligent Mobilization' under theme 1, 'Innovative State and Public Services' under theme 4, 'Information and Communication Technologies for Inclusion' and 'Open Innovation for Internet Based Services' under theme 5. In these projects, the state, municipalities, private sector and non-governmental organizations (NGOs) will also make applications for smart cities together (Alkan, 2014).

According to 'Municipalities Smart City Applications Survey' performed with 40 municipalities in February 2013, smart city applications come to the forefront in the field of urban services and transportation in the cities, smart city applications in the field of energy and water were realized by less number of municipalities. During urban transformation process, smart city applications present important opportunities. Ensuring urban transformation design, management and the participation of the citizens in the process, being used of the innovative applications provides time and cost saving. In this scope, some measurements such as forming the smart city substructures, giving place for smart building applications in the buildings to be newly built contribute to the increase in life standards. (2014-2018 Kalkınma Bakanlığı., 2014) (Figure 4).

About smart city implementations in Turkey, Istanbul is leading and there is also some work about this subject in Ankara (Table 6; Yilmaz, 2015).

Istanbul Fatih Municipality goes one step further and incorporates the implementation of augmented reality into Smart City projects. According to this application, when the image of any building in Fatih Municipality is photographed and sent to the related service center via 3G-4G communication technology, the existing information about that building can be transferred to the user immediately from the information center (*Akıllı Kent Masabaşı Araştırması*. Kamu, 2016b).

Figure 4. Smart city applications used in Turkish municipalities, 2013

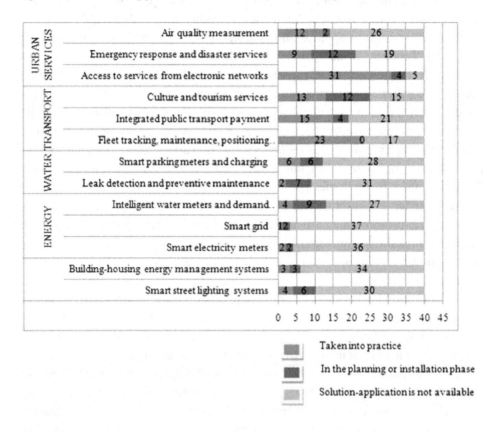

Table 6. Smart city solutions in Turkey (Istanbul and Ankara)

ISTANBUL	ANKARA
✓ Smart Counters: Bedaş, Ayedaş ✓ EU Smart Space: Smart Buildings, Fatih sport complex ✓ ISKI: Drinking water distribution management ✓ IBB: Traffic control center ✓ ISPARK: Smart park pricing and automatic parks ✓ Cart: Integrated cost payment for different transportation vehicles ✓ ISBANK: Smart containger, mobese and digital library	✓ ASKI, The system for obtaining administrative audit and data, ✓ EGO, EGO In the Pocket, Smart Stop, ✓ In-vehicle Passenger Information and Cam Systems ✓ Smart Structures, Industry Park ✓ Automatic Fire Station Command Center

Table 7. Smart city project examples in Turkey

City	Project Framework	Project Resource	Project Process	The Stage of Project Development
Yalova	E-Municipality, Smart City	Local resources	In progress	The Silicon Valley Project is not yet completed
Fatih	E-Municipality, Smart City	Local resources	In progress	GIS and numerical mapping work of Fatih District was completed, Smart City project implementation group established
Kadıköy	E-Municipality, Smart City	Local resources	In progress	GIS and numerical mapping work of Kadıköy District was completed
Beyoglu	E-Municipality, Smart City	Local resources	In progress	GIS and numerical mapping work of Beyoglu District was completed, System integration is being carried out with the e-Municipal applications
Izmir	E-Municipality, Smart City	Local resources	In progress	GIS and numerical mapping work of İzmir continues and the 3D identification work continues
Ankara	E-Municipality, Smart City	Local resources	In progress	GIS and numerical mapping work of Ankara continues
Bursa	E-Municipality, Smart City	Local resources	In progress	GIS and numerical mapping work of Bursa continues

(Akıllı Kent Masabaşı Araştırması. Kamu, *2016b).*

GIS substructure that forms basis for smart city planning has been realized by a small number of municipalities in our country. According to e-State Survey performed by the Ministry of Internal Affairs in 2011 on 90 percent of all municipalities, GIS work has been completed only in 3 percent of the municipalities and in 14 percent of them they have been continuing partly. The need of the municipalities that have not technical capacity to form GIS substructure in being supported about this matter has been observed. It is among smart street illumination systems; building/house energy management systems; smart electric counters; smart electricity network; smart water counters and demand management; the detection of the leakages and preventive care; smart park meters and pricing; fleet monitoring, maintenance, location identification services; integration public transportation services; culture and tourism payment, access to the services through electronic channels, emergency intervention and disaster services, air quality monitoring and transportation smart city applications (2014-2018 Kalkınma Bakanlığı., 2014).

National and international smart city organizations The 'Smart City' organization that stands out in the world is the United States (USA) based 'Smart City Council'. The Smart Cities Council envisions 'a world in which sustainable cities are created for high-quality living and quality business environments intelligently designed with digital technology' as a vision. The Council of Smart Cities has also taken on the task of 'advising on the development and progress of the sector, encouraging decision makers to act for smart and sustainable cities, and contributing to the stakeholders' success'. There are many global companies and NGOs that support the issue and the membership of the Council. These include "AT&T, Cisco, GE, Verizon, Daimler, Microsoft, Schneider Electric, IBM, MasterCard, S&C Electric, Qualcomm, ABB, Intel, Siemens, Universitat Autònoma de Barcelona, National Governors Association, EDF), The Climate Group, Boyd Cohen, Universidad del Desarrollo, Pedro Ortiz, Senior Urban Consultant, World Bank, Portland Development Commission, US Green Building Council (USGBC), Institute for Sustainable Communities, CompTIA, Energy Future Coalition" (Smart Cities Council, n.d.).

In Turkey, there is still no civil organization related to smart cities. However, the Ministry of Environment and Urbanization and the Ministry of Energy and Natural Resources are contributing in terms of renewable energies to the related public and stakeholders. The Union of the Municipalities of Turkey, as an NGO, has been continuing as a state institution as the Local Authorities Association and contributing to the activities related to the issue. In the vision of the future, the Ministry of Development, in its document 'Smart City'; today's solutions to solve the problems of the cities and to increase the quality of life of the people living in the cities gain importance and in many cities around the world" It is being applied quickly. The 'Transformation Leaders Board', which was established in 2003 with the Prime Ministry Circular No. 48 and the e-Transformation Turkey Executive Board and the Prime Ministry Circular No. 7 in 2007 as the closest addressee in Turkey. However, there is still a need for an NGO that will directly address the issue for the future of smart cities in the country (Alkan, 2014).

Intelligent Transportation Systems (ITS) in Turkey

In recent years, the environmental impacts of transportation systems have reached alarming dimensions. Nevertheless, the use of motor vehicles continues to increase and as a result, traffic pollution is increasing. In particular, measures must be taken to reduce carbon dioxide (CO_2) and nitrogen oxide (NOx) emissions. ITS help reduce vehicle emissions due to benefits such as reducing traffic congestion and allowing private motorists to move to public transport. In addition, intelligent transport systems allow pricing to be applied where drivers or vehicles are being used at times of high pollution or when the release of such emissions is high. Thus, these systems will help to reduce environmental impacts.

If the rate of urbanization of developed industrial countries such as the United Kingdom, Sweden and Japan is still increasing, and if the rate of Turkey's urbanization is already low compared to these countries, It can be predicted that this ratio will continue to increase for a long time in Turkey. The suburban phenomenon, which gained importance with the acceleration of urbanization, gave rise to an accelerated rise in demand for mobility within the city, which caused every city that became a center of attraction to face a traffic jam. The beginning of separation of working and residential areas necessitates the use of motor vehicles between the home and the workplace, especially in the big cities. As the attractiveness of the suburbs which far from the noise of the city center increases, the demand for personal vehicles also increases (UDHB, 2014).

In recent years, significant amounts of investments have been made to the transport and communications sector in Turkey. In this sense, important developments have also been made regarding the widespread use of intelligent transport systems.

It has been decided to prepare a strategy document in order to bring the Smart Transportation Systems to a planned and systematic structure in Turkey. Intelligent Transportation Systems Strategy Document is prepared to create a roadmap to increase the use of ITS on highways. This study was prepared with the participation and contributions of the General Directorate of Highways, the Ministry of Development, the Ministry of Interior, Metropolitan Municipalities, universities, related private sector and non-governmental organizations (NGO) in with coordination of the Ministry of Transport, Maritime and Communication.

Table 8. Percentage of urban population to total population (%) (Data Source: World Bank, 2017)

Country	2006	2007	2008	2009	2010	2011
USA	81,01	81,29	81,57	81,86	82,14	82,38
Germany	73,44	73,53	73,63	73,72	73,81	73,94
Austria	66,71	66,90	67,08	67,27	67,45	67,66
France	82,28	83,02	83,75	84,49	85,22	85,74
South Korea	81,66	81,98	82,29	82,61	82,93	83,20
Netherlands	80,68	81,20	81,71	82,23	82,74	83,13
Britain	79,10	79,20	79,30	79,40	79,50	79,63
Swedish	84,46	84,61	84,76	84,90	85,05	85,20
Japan	86,89	87,80	88,71	89,62	90,54	91,13
Norway	77,81	78,13	78,45	78,77	79,10	79,37
Turkey	67,57	68,29	69,02	69,75	70,48	71,40
OECD	78,02	78,36	78,70	79,04	79,38	79,68
World	49,55	50,05	50,54	51,03	51,52	51,99

While there are a large number of direct and indirect benefits expected from intelligent transportation systems, the basic expectations can be summarized as follows:

1. Reducing traffic accidents, increasing the safety and security of roads, drivers and pedestrians,
2. Improving the performance of the transport system and reducing traffic congestion,
3. Monitoring, directing and real time management of traffic,
4. Optimization of transport times and contribution to the economy by reducing transportation costs,
5. Increasing service quality and productivity,
6. Increasing personal mobility and comfort,
7. Minimizing the damage to the environment and saving energy (UDHB, 2014).

Research has shown that applications of intelligent transport systems always precede legislation and politics. Firstly, technologies were developed, then applications using this technology were invented, but only after those applications started to become widespread and localized, the usage policy, strategy and legislation of the application were put forward. For this reason, some applications can not be clearly classified according to any of the several classification methods available. Although not a generally accepted classification of intelligent transportation systems, applications can be classified under the following headings:

1. Passenger Information Systems,
2. Traffic Management Systems,
3. Public Transportation Systems,
4. Electronic Payment Systems,
5. Vehicle and Fleet Management Systems,
6. Driver Support and Security Systems,
7. Accident and Emergency Systems,
8. Railways Operation and Management.

Within the scope of National Transportation Portal (Turkey), according to the transportation between two points and individual or public transport options, it is possible to make travel planning. Transactions can be performed, including ticketing via the portal. The portal also includes components such as road status, accident information, weather, radio, and so on. It has components such as mobile application, 3-dimensional street view, travel with more than one type of transportation, traffic education for children with four language options (UDHB, 2014).

Figure 5. National transportation portal (Turkey) (Ulusal Ulastırma Portali)

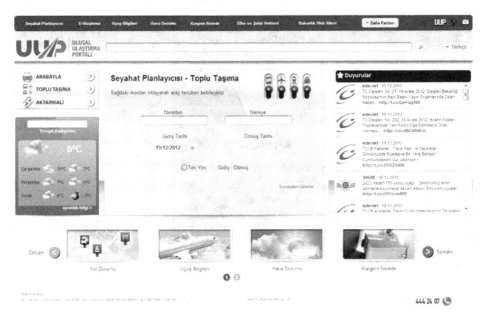

IBB Mobile Traffic (Turkey) IBB Mobile Traffic is a service that instantly shows the traffic situation in Istanbul with its density map and live camera images to its users instantly. This service has been downloaded approximately 1.500.000 times in the last 4 years and the current active user is about 1 million. With IBB Mobile Traffic (UDHB, 2014);

- Instant density information of traffic can be reached in Istanbul,
- It can be connected to all the traffic cameras at different points of Istanbul and get images instantly,
- Frequently used cameras are added to the favorite and these images can be accessed with shortcut keys,
- With shortcut keys; the location can be seen, the location of the bridge can be located according to Istanbul, regardless of the location, can be made to switch to users favorite camera,
- Only the density map can be reached by closing the cameras,
- Filtered details can be displayed on the map by performing a location search on the map.

Traffic control systems detect instantaneous intensity in traffic and regulate signal durations accordingly. This system is used by Istanbul Metropolitan Municipality and some other municipalities. With this system, many benefits can be gained such

Figure 6. IBB mobile traffic service

as optimization of signaling times and road capacities, reduction of air pollution level and expenses such as fuel, spare parts, shortening travel and waiting times (UDHB, 2014).

Smart City Policies in Turkey

In Turkey, national and local governments have begun to develop strategies for smart city solutions. Smart city practices are shaped by top policy documents, ministry strategies and studies, initiatives of local governments and technology providers and platforms created by non-governmental organizations. With the determination of the smart city needs and taking the steps in this direction, smart city applications started to be dealt with in different top policy documents created by the government and initiatives have been started in Turkey. The strategic plans of the relevant ministries as well as the high-level policy documents are listed in Table 9 (*Akıllı şehir yol haritası Vodafone.*, 2016a).

A hybrid approach of 'top-down' and 'bottom-up' policies has been adopted within the Information Society Strategy to support the transformation of Turkish cities into smart cities (2015-2018 Kalkınma Bakanlığı., 2015). Smart city policies in the New Information Society Strategy of Turkey are indicated in Table 10, Table 11 and Table 12.

According to the field study done, the top policy documents examined and the digitization index findings in Turkey (*Akıllı şehir yol haritası Vodafone.*, 2016a):

Table 9. The top policy documents of initiatives on smart cities in Turkey

The Tenth Development Plan (2014-2018)	• Information Society • Transportation • Health • Water Management • Energy • Building
Medium-Term Programme (2015-2017)	• Urbanization • Innovation • Energy • Environment
National E-Government Strategy and Action Plan	• E-Government Vision • Core Values • Strategic Objectives and Targets
KENTGES-Integrated Urban Development Strategy and Action Plan	• Productivity • Water Management • Waste Management • Life Quality • Sustainability
National Science Technology and Innovation Strategy	• Research and Development (R&D) • International Cooperation • Qualified Human Resource
Information Society Strategy and Action Plan	• Technology • Information systems • Financing Proposal • Qualified Human Resource
Energy Efficiency Strategy Document	• Smart Grids • Energy • Productivity
Traffic Safety Action Plan	• Transportation • Road Safety
Transportation and Communication Strategy Target 2023	• Smart Communication Networks • Local Software • Carrying Trade • Transportation
National Climate Change and Strategy Document (2010-2023)	• Environment • Climate • Energy • Transportation • Waste Management • Greenhouse Gas (CO2) Release

Table 10. Smart cities in the new information society strategy of Turkey; smart urban program development (2015-2018)

Smart Urban Program Development (2015-2018)	
Policy	Necessary precautions will be taken to establish smart cities. For this purpose, strategies and targets will be determined and policies will be set up for the integration of working principles and the required governance models. Smart urban applications will be given priority in metropolitan cities and areas covered by urban transformation. A road map will be created for this purpose. Intelligent transport systems will be developed and coordination between the different applications of these institutions will be ensured.
Definition	Strategies and objectives will be determined for the realization and dissemination of smart urban solutions. In this context, this strategy will determine the areas to focus on in smart city solutions, objectives, stakeholder engagement methods and the necessary financing model. In this context, firstly the standards of smart urban practices will be revealed and various R&D studies will be supported by taking these standards into consideration. In addition, commercialization of developed technological products will enable the efficient use of public procurement and pilot applications in the living laboratory. For this purpose, smart urban applications will be supported with public resources, especially the resources provided by local administrations of ILBANK.
Responsible and collaborative organizations	Ministry o f Environment And Urbanization, Ministry of Development, Ministry of Health, Ministry of Transport, Maritime Affairs and Communications, Ministry of Energy and Natural Resources, Ministry of Forest and Water Affairs, Ministry of Science, Industry and Technology, Ministry of Interior, Ministry of Finance, ILBANK, TUBITAK, TSE, TOKI, Disaster and Emergency Management, Development Agencies, Local Governments, Universities, NGOs.

Table 11. Smart Cities in the new information society strategy of Turkey; supporting smart applications (2015-2018)

Supporting Smart Applications (2015-2018)	
Policy	Call-based support will be provided for the development of smart applications by using public data generated by central agencies and local governments.
Definition	High value-added services (especially mobile applications and geographical services) will be developed and also these services will be provided for use by citizens. In this context, the projects will be supported by the private sector and universities that in areas such as health, transportation, building, energy, disaster and water management within the scope of the priorities set by the Development Agencies. Thus, private sector and university creativity and public data will be brought together to create innovative solutions with added value.
Responsible and collaborative organizations	Ministry o f Environment and Urbanization, Ministry of Development, Ministry of Health, Ministry of Transport, Maritime Affairs and Communications, Ministry of Energy and Natural Resources, Ministry of Forest and Water Affairs, Disaster and Emergency Management, Development Agencies, Local Governments, Universities, Development Agencies.

Table 12. Smart Cities in the new information society strategy of Turkey; development of living laboratories program

Development of Living Laboratories Program	
Policy	The living laboratories approach will be adopted to the development and commercialization of high-tech products. In this context, the environment in which the citizens are involved will be passed on in the innovation process.
Definition	Within the scope of the program, the concept of living laboratory which will accelerate the commercialization of the developed products by participating in the innovation process of the citizens will be defined. Also, the principles and methods of living laboratories and the coordination between different laboratories will be determined.
Responsible and collaborative organizations	Ministry of Science, Industry and Technology, Ministry of Development, Revenue Administration, TUBITAK, Development Agencies, Local Governments, Universities, NGOs.

- Local governments are highly aware of the smart city in Turkey. Vision, planning and team building are being carried out. According to the survey results, 17 of the 19 metropolitan municipalities have a vision of a smart city. A smart city plan has been prepared in 15 of the 17 municipalities. However, it is seen that the prepared plans are in a fragmented structure in the form of transportation or energy master plan.

- It has been seen that there is a remarkable awareness, interest and intention towards becoming a smart city and top policies also support this. The perception that a holistic approach is needed to be a smart city has begun to emerge.

- Smart city transformation is seen as an important opportunity for actors in national and local scale. Smart city solutions are a necessity for local governments and they have focused on finding solutions to problems such as traffic, transportation, citizen satisfaction, environmental management and security that come with increasing population.

- As the municipal scale changes, the priorities of the problems are also changing. The concept of making a difference with technology is dominant in service provision. In other words, local governments see technology as the most critical solution in their service area.

- It is envisaged that the smart cities' contribution to Turkey's gross domestic product would be between 6 and 7,5 billion dollar annually if smart city programs are designed and implemented in a country-wide manner. When successful examples of applications are evaluated, it is possible to save 20% of energy if 30 big cities are turned into smart cities in Turkey.

The steps to be taken in order to achieve economic and social sustainability in Turkey are listed as in the following subheadings (*Akıllı şehir yol haritası Vodafone.*, 2016a).

To Ensure Social Sustainability

- In a smart city that aims to create a sustainable and healthy living environment, all segments of the society should be included in the system and this experience should be provided,
- Increasing the quality of life and delivering services to all segments of society,
- Achieving the benefit of all residents of the city from smart city solutions and improving quality of life for everyone,
- Making cost-effective social assistance programs for disadvantaged groups better at lower cost,
- The use of technological tools to make disadvantaged groups more accessible to government services,
- Developing communication channels with excluded groups in order to understand their needs and alleviate living difficulties,
- Development of support programs to enable disadvantaged citizens to participate in projects due to social exclusion,
- Projects that both the private and the public sector can benefit from should be developed. The use of private sector financing and resources should be encouraged,
- A solution to the financing constraints of the municipalities should be provided. Sustainable economic growth should be supported through private sector investments,
- Communication and cooperation channels between public and private sector should be established,
- Projects that can offer specific solutions to the needs of cities and different institutions should be developed.

To Ensure Economic Sustainability

- The financing constraints of the municipalities must be resolved and sustainable economic growth should be promoted through private sector investments,
- The appropriateness of private sector activities and financial plan should be controlled by strong institutions and control mechanisms should be developed,
- A market should be established where large companies and small enterprises can compete on equal terms,

- Many projects must be run under a single authority.

In the formation of smart cities, the applications such as mutual movement with the public, being sustainable, the studies of identifying techniques and areas to be used and receiving feedback with the surveys about the subjects, smart counters, public transportation payment types, the prevention of carbon release in transportation systems, bringing the electricity to the forefront as fuel and online monitoring the transportation vehicles, etc. are discussed before public opinion and turned into action. On the information side of the subject, it is emphasized that especially to how big extent the costs of non-informativeness activities are. The detection of a policy with the name of smart cities and making both its sanctions and monitoring, meeting the financial needs in the fastest way, the opportunities of making use of EU funds, national agencies and the other funds must be searched. It is required to act commonly with the district municipalities especially in the scope of metropolitan city and enact law that will ensure the central sanctions to orient giving support to the municipalities at once and to identify smart city components at once and giving training on this matter at once. It should be done common work with relevant organizations of the public under the name of interoperability about the subject (Ministry of Interior, Affairs, General Directorate of Local Authorities, Environment and Urban Planning Ministry, the Ministry of Development, Bank of Provinces General Directorate etc.), non-governmental organizations (NGOs), private sector collaboration and the Municipalities Union and they must monitor the process together. In electronic media where local administrations can share information with each other, the idea of forming a platform under the name of "Smart Cities Cooperation Platform" must be brought to the agenda. It can be a good step to discuss the SMARTCITIES program of EU as an adaptable model even if it does not comply with our country one by one. First of all, 30 metropolitan cities will be discussed and in the framework of a model to be prepared, it can be passed to the applications and in the next stage it can be disseminated (Cerci, 2015).

Smart city transformations are a process that requires long-term, investment and mutual understanding of many different sides. Progress should be made with the right technology and the targeted effect must be planned at every step. The motto of a healthy smart transformation can be summarized as planning, reconciliation and ownership (*Akıllı şehir yol haritası Vodafone.*, 2016a).

CONCLUSION

In the solution of urban problems in Turkey, smart city implementations have an important potential. Many cities have realized/are realizing smart applications (especially in the fields of transportation and urban services area). However, the number of applications in the field of energy and water management is extremely limited. The biggest obstacle for the realization of smart city applications is the insufficiency in human resources and the problems to access financial resources. Besides, the deficiencies in the substructure of GIS constitute a significant obstacle.

For the realization of smart city implementations in a holistic way, it is needed to strong financial resources and a strong long-term foresight. Smart city solutions are required to be prioritized according to the preferences of the citizens. Instead of independent or small-scaled pilot implementations, the projects having inter-sectoral integration must be preferred. By this way, the problems in data compliance among the solutions will be eliminated. For the realization of it in an integrated and holistic way, serious financial resources and strong long-term foresight are required. State and private sector financing have great importance in realizing smart city solutions. In all stages of the applications, the participation of the public must be ensured.

The recommendation is put forward for the progress of smart urbanization in Turkey is summarized as follows (*Akıllı Kent Masabaşı Araştırması*. Kamu, 2016b).

Funds need to be set up to realize smart city projects. These funds must be used for the projects to be realized in cooperation with NGOs, universities and municipalities in the specified standards. Accordingly, as one of the success factors of the projects, it is important to start working towards increasing cooperation and synergies among the institutions. It is essential to utilize the information and data analysis for current smart city applications. However, while information technology is being used in these projects, it is necessary to use the data privacy policies of related institutions. Given the lack of knowledge and experience on the subject of smart cities, the prominence of a competent human resources issue arises. The inadequacy of human resources in smart city projects must be overcome. It is proposed to design products and services that are citizens oriented and target social innovation. It can be incorporated into city decision making mechanisms through citizen's smart city platforms.

It is proposed that a vision of a national mechanism for smart cities to be found in local government and related organizations on issues such as national strategy, urban transformation, standards, cooperation, interoperability of systems, data exchange between institutions is suggested. Especially in the developing regions, the expertise and capacity of ICT in the scope of smart cities to organize events and to contribute to the elimination of human resources in smart city projects. It is imperative that Turkey should review the successful smart city models in different regions of the world and develop application models for Turkish cities that are being converted or newly established and prepare separate road maps for different cities. It is important to initiate cooperation activities so that successful smart city applications can be implemented in other provinces, districts and institutions in our country. Finally, it is necessary to evaluate the concepts and application models by applying them on a small scale and try to enlarge the time scales if they are successful.

REFERENCES

2014-2018 Kalkınma Bakanlığı. (2014). *Bilgi toplumu stratejisi ve eylem planı taslağı.* Retrieved from http://www.bilgitoplumustratejisi.org/tr/doc/8a94819842e 4657b01464d5025b80002

2015-2018 Kalkınma Bakanlığı. (2015). *Bilgi toplumu stratejisi ve eylem planı.* Retrieved March 24, 2017, from http://www.bilgitoplumustratejisi.org/tr/doc/8a94 81984680deca014bea4232490005

Akgül, M. (2013). Kentlerin e-dönüşümü akıllı kentler. *Kalkınmada Anahtar Verimlilik Dergisi, 291.* Retrieved March 12, 2016, from https://anahtar.sanayi.gov. tr/tr/news/kentlerin-e-donusumu-akilli-kentler/416

Akıllı Kent Masabaşı Araştırması. (2016b). Teknoloji Platformu XSights. Retrieved March 12, 2017, from http://www.akillisehirler.org/wp-content/uploads/2016/08/ Xsights-Ak%C4%B1ll%C4%B1-Kentler-Masa-Ba%C5%9F%C4%B1- %C3%87al%C4%B1%C5%9Fmas%C4%B1.pdf

Akıllı şehir yol haritası Vodafone. (2016a). Türkiye Bilişim Derneği Katkılarıyla. Retrieved from https://www.sehirsizin.com/Documents/Deloitte-Vodafone-Akilli- Sehir-Yol-Haritasi.pdf

Alkan, T. (2014). Akıllı şehirler ve hybrid bilişim (ICT) ağı. *TBD Ankara Bilişim Kongresi.* Retrieved May 30, 2016, from http://www.bilisimdergisi.org/s182/pdf/70- 77.pdf

Amsterdam City. (n.d.). Retrieved May 12, 2016, from http://amsterdamsmartcity. com/

Ayber, A. H. (2016). *Akıllı Kentler.* Smart Cities.

Belissent, J. (2010). *A Tale of Two Cities: Two Approahces to Making Cities Smarter, Part III.* Retrieved May 12, 2016, from https://go.forrester.com/blogs/10- 03-15-a_tale_of_two_cities_two_approaches_to_making_cities_smarter_part_iii/

Carriero, D. (2015). United smart cities: Smart urban solutions for transition and developing countries. In *9th session of the WPLA.* UNECE. https://www.unece.org/ fileadmin/DAM/hlm/wpla/sessions/9th_session/day_2_presentations/13_Carriero_ WPLA9_day2.pdf

Cerci, N. (2015). Akıllı kentler bir ihtiyaç ve geleceğe dönük bir yaşam modelidir. *Bilişim Dergisi, 43*(172). Retrieved May 12, 2016, from http://www.bilisimdergisi. org/s172/FLASH/index.html

Digital Age. (2014). *Dünyadan Akıllı Şehir Örnekleri*. Retrieved April 07, 2017, from http://digitalage.com.tr/dunyadan-akilli-sehir-ornekleri/

Jain, C. (2015). *UTC India-Student Essay-Inclusive Smart City*. Retrieved March 02, 2017, from http://utc.niua.org/utc-india-student-essay-inclusive-smart-city/

Kalkınma Bakanlığı, T. C. (2013). *Bilgi toplumu stratejisinin yenilenmesi projesi bilgi ve iletişim teknolojileri destekli yenilikçi çözümler ekseni küresel eğilimler ve ülke incelemeleri raporu*. Retrieved March 24, 2017, from www.bilgitoplumustratejisi.org

Kumar, P. (2015). *What's the Real Mean of "Smart City"?* Retrieved March 12, 2017, from http://www.smartcitiesprojects.com/whats-the-real-mean-of-smart-city/

Lugaric, L., Krajcar, S., & Simic, Z. (2010). Smart City - Platform for Emergent Phenomena Power System Testbed Simulator. *Innovative Smart Grid Technologies Conference*. IEEE Xplore. Retrieved March 30, 2017, http://ieeexplore.ieee.org/stamp/stamp.jsp?arnumber=5638890

McKinsey and Company (2010). *A lean approach to energy efficiency*. Author.

Nijkamp, P., & Pepping, G. (1998). A Meta-analytical evaluation of sustainable city initiatives. *Urban Studies (Edinburgh, Scotland)*, *35*(9), 1481–1500. doi:10.1080/0042098984240

Sınmaz, S. (2013). The concept of "smart settlement" and basic principles in the framework of new developing planning approaches. *Megaron*, *8*(2), 76–86. doi:10.5505/megaron.2013.35220

Smart Cities Council. (n.d.). Retrieved March 28, 2017, from http://smartcitiescouncil.com/

Smart Cities in Europe. (2016). Retrieved May 14, 2016, from, ftp://zappa.ubvu.vu.nl/20090048.pdf

Smart Cities Mission. (2016). Retrieved May 30, 2016, from http://smartcities.gov.in/writereaddata/SmartCityGuidelines.pdf

Smart Cities Readiness Guide. (2015). *The planning manual for building tomorrow's cities today*. Smart Cities Council. Retrieved February 12, 2017, from http://www.uraia.org/OlalaCMS4/files/574_arquivoB.pdf

Songdo, I. B. D. (2015). Retrieved May 14, 2016, from http://www.songdo.com

UDHB. (2014). *Ulusal akıllı ulaşım sistemleri strateji belgesi ve eylem planı*. Ankara: T.C. Ulaştırma, Denizcilik ve Haberleşme Bakanlığı.

Ulusal Ulastiurma Portali. (n.d.). Retrieved March 28, 2017, from www.ulasim.gov.tr

UN. (1987). *Our common future report.* UN Documents. http://www.un-documents. net/wced-ocf.htm

UN. (2015). *World population prospects the 2015 revision.* New York: Department of Economic and Social Affairs.

UN. (2016). *World urbanization prospects.* Department of Economic and Social Affairs, Population Division.

UNESCO-MOST. (1996). Conference report on sustainability as a social science concept. Frankfurt: Author.

Van Geenhuisan, M., & Nijkamp, P. (1994). Sürdürülebilir kenti nasıl planlamalı? *Toplum ve Bilim Dergisi, 64*(65), 129–140.

World Bank. (2017). Retrieved May 13, 2017, from http://www.worldbank.org

Yilmaz, E. (2016). *Yeni bir hikaye Konya akıllı şehir olabilir mi?* Retrieved May 30, 2016, from http://www.kto.org.tr/d/file/yeni-bir-hikaye-konya-akilli-sehir-olabilir-mi---emre-yilmaz.20160222152345.pdf

Compilation of References

2014-2018 Kalkınma Bakanlığı. (2014). *Bilgi toplumu stratejisi ve eylem planı taslağı.* Retrieved from http://www.bilgitoplumustratejisi.org/tr/doc/8a94819842e 4657b01464d5025b80002

2015-2018 Kalkınma Bakanlığı. (2015). *Bilgi toplumu stratejisi ve eylem planı.* Retrieved March 24, 2017, from http://www.bilgitoplumustratejisi.org/tr/doc/8a94 81984680deca014bea4232490005

Ahlemann, F., El Arbi, F., Kaiser, M., & Heck, A. (2013). A process framework for theoretically grounded prescriptive research in the project management field. *International Journal of Project Management, 31*(1), 43–56. doi:10.1016/j. ijproman.2012.03.008

Ahmad, S., Liu, M., & Wu, Y. (2009). *Congestion games with resource reuse and applications in spectrum sharing.* CoRR, abs/0910.4214

Ahmed, J., Rogge, R., Kline, W., Bunch, R., Mason, T., & Wollowski, M. (2014). The innovation canvas: An instructor's guide. In *121st ASEE Annual Conference and Exposition* (pp. 1-12). Indianapolis, IN: American Society for Engineering Education.

Akgül, M. (2013). Kentlerin e-dönüşümü akıllı kentler. *Kalkınmada Anahtar Verimlilik Dergisi, 291.* Retrieved March 12, 2016, from https://anahtar.sanayi.gov. tr/tr/news/kentlerin-e-donusumu-akilli-kentler/416

Akıllı Kent Masabaşı Araştırması. (2016b). Teknoloji Platformu XSights. Retrieved March 12, 2017, from http://www.akillisehirler.org/wp-content/uploads/2016/08/Xsights-Ak%C4%B1ll%C4%B1-Kentler-Masa-Ba%C5%9F%C4%B1-%C3%87al%C4%B1%C5%9Fmas%C4%B1.pdf

Akıllı şehir yol haritası Vodafone. (2016a). Türkiye Bilişim Derneği Katkılarıyla. Retrieved from https://www.sehirsizin.com/Documents/Deloitte-Vodafone-Akilli-Sehir-Yol-Haritasi.pdf

Al Nuaimi, E., Al Neyadi, H., Mohamed, N., & Al-Jaroodi, J. (2015). Applications of big data to smart cities. *Journal of Internet Services and Applications*, *6*(1), 25. doi:10.118613174-015-0041-5

Alabdulkarim, L., & Lukszo, Z. (2008). Information security assurance in critical infrastructures: Smart Metering case. *2008 1st International Conference on Infrastructure Systems and Services: Building Networks for a Brighter Future, INFRA 2008*. 10.1109/INFRA.2008.5439670

Alawadhi, S., Aldama-Nalda, A., Chourabi, H., Gil-Garcia, J. R., Leung, S., Mellouli, S., . . . Walker, S. (2012). September. Building understanding of smart city initiatives. In *International Conference on Electronic Government* (pp. 40-53). Springer Berlin Heidelberg. 10.1007/978-3-642-33489-4_4

Albert, A., & Rajagopal, R. (2013). Smart meter driven segmentation: What your consumption says about you. *IEEE Transactions on Power Systems*, *28*(4), 4019–4030. doi:10.1109/TPWRS.2013.2266122

Albino, V., Berardi, U., & Dangelico, R. M. (2015). Smart cities: Definitions, dimensions, performance, and initiatives. *Journal of Urban Technology*, *22*(1), 3–21. doi:10.1080/10630732.2014.942092

Alkan, T. (2014). Akıllı şehirler ve hybrid bilişim (ICT) ağı. *TBD Ankara Bilişim Kongresi*. Retrieved May 30, 2016, from http://www.bilisimdergisi.org/s182/pdf/70-77.pdf

Alkon, A., & Agyeman, J. (2011). *Cultivating food justice: race, class, and sustainability*. Cambridge, MA: The MIT Press.

Allianz. (n.d.). KIT Industry 4.0 Collaboration Lab. *Allianz Industrie 4.0*. Retrieved from: https://www.i40-bw.de/de/100orte/imi-am-kit/

Almadhor, A. (2016). A Fog Computing based Smart Grid Cloud Data Security. *International Journal of Applied Information Systems*, *10*(6), 1–6. doi:10.5120/ijais2016451515

Alzate, C., & Sinn, M. (2013). Improved electricity load forecasting via kernel spectral clustering of smart meters. *Proceedings - IEEE International Conference on Data Mining, ICDM*, 943–948. 10.1109/ICDM.2013.144

Amadi, L., & Imoh-Ita, I. (2017). Intellectual capital and environmental sustainability measurement nexus: a review of the literature. *Int. J. Learning and Intellectual Capital, 14*(2), 154-176.

Amadi, L., & Igwe, P. (2016). Maximizing the Eco Tourism Potentials of the Wetland Regions through Sustainable Environmental Consumption: A Case of the Niger Delta, Nigeria. *The Journal of Social Sciences Research, 2*(1), 13–22.

Aman, S., Simmhan, Y., & Prasanna, V. K. (2011). Improving energy use forecast for campus micro-grids using indirect indicators. *Proceedings - IEEE International Conference on Data Mining, ICDM*, 389–397. 10.1109/ICDMW.2011.95

Amsterdam City. (n.d.). Retrieved May 12, 2016, from http://amsterdamsmartcity.com/

Anthopoulos, L., & Vakali, A. (2012). Urban planning and smart cities: Interrelations and reciprocities. The Future Internet, 178-189.

Aoun, C., (2013). *The smart city cornerstone: Urban efficiency*. Schneider Electric.

Ardito, L., Procaccianti, G., Menga, G., & Morisio, M. (2013). Smart grid technologies in Europe: An overview. *Energies, 6*(1), 251–281. doi:10.3390/en6010251

Arora, S., & Taylor, J. W. (2016). Forecasting electricity smart meter data using conditional kernel density estimation. *OMEGA-International Journal of Management Science, 59*(A, SI), 47–59. 10.1016/j.omega.2014.08.008

Artto, K., & Kujala, J. (2008). Project business as a reseach field. *International Journal of Managing Projects in Busines, 1*(4), 469–497. doi:10.1108/17538370810906219

Asbery, C. W., Jiao, X., & Liao, Y. (2016). Implementation guidance of smart grid communication. *Proceedings of 2016 North American Power Symposium (NAPS)*, 1-6.

Ashton, K. (2009). That "Internet of Things" thing. *RFID Journal, 22*, 97–114.

Asif, M., Searcy, C., Zutshi, A., & Fisscher, O. (2013). An integrated management systems approach to corporate social responsibility. *Journal of Cleaner Production, 56*, 7–17. doi:10.1016/j.jclepro.2011.10.034

Augusto, J., Callaghan, V., Cook, D., Kameas, A., & Satoh, I. (2013). Intellignet Environments: a manifesto. *Human-centric Computing and Information Sciences, 3*(12).

Ayber, A. H. (2016). *Akıllı Kentler.* Smart Cities.

Baimel, D., Tapuchi, S., & Baimel, N. (2016). Smart grid communication technologies-overview, research challenges and opportunities. *Proceedings of 2016 International Symposium on Power Electronics, Electrical Drives, Automation and Motion (SPEEDAM)*, 116-120. 10.1109/SPEEDAM.2016.7526014

Ballon, P., Glidden, J., Kranas, P., Menychtas, A., Ruston, S., & Van Der Graaf, S. (2011), October. Is there a need for a cloud platform for European smart cities? In *eChallenges e-2011 Conference Proceedings, IIMC International Information Management Corporation* (pp. 1-7). Academic Press.

Bari, A., Jiang, J., Saad, W., & Jaekel, A. (2014). Challenges in the smart grid applications: An overview. *International Journal of Distributed Sensor Networks, 10*(2), 974682. doi:10.1155/2014/974682

Barnett, T. (2004). *The Pentagon's New Map.* Putnam Publishing Group.

Barrionuevo, J. M., Berrone, P., & Ricart, J. E. (2012). Smart cities, sustainable progress. *IESE Insight, 14*(14), 50–57. doi:10.15581/002.ART-2152

Barr, S., Gilg, A., & Ford, N. (2005). The household energy gap: Examining the divide between habitual- and purchase related conservation behaviors. *Energy Policy, 33*(11), 1425–1444. doi:10.1016/j.enpol.2003.12.016

Bartoli, A., Hernández-Serrano, J., Soriano, M., Dohler, M., Kountouris, A., & Barthel, D. (2011). Security and privacy in your smart city. In *Proceedings of the Barcelona smart cities congress* (pp. 1-6). Academic Press.

Beckel, C., Sadamori, L., Staake, T., & Santini, S. (2014). Revealing household characteristics from smart meter data. *Energy, 78*(SI), 397–410. 10.1016/j.energy.2014.10.025

Beckel, C., Sadamori, L., & Santini, S. (2013). Automatic socio-economic classification of households using electricity consumption data. *Journal of Economic Psychology, 3*(3–4), 75–86. doi:10.1016/0167-4870(83)90006-5

Belissent, J. (2010). *A Tale of Two Cities: Two Approahces to Making Cities Smarter, Part III*. Retrieved May 12, 2016, from https://go.forrester.com/blogs/10-03-15-a_tale_of_two_cities_two_approaches_to_making_cities_smarter_part_iii/

Bélissent, J., (2010). *Getting clever about smart cities: New opportunities require new business models*. Academic Press.

Bennis, W. (2013). Leadership in a Digital World: Embracing Transparency and Adaptive Capacity. *Management Information Systems Quarterly, 37*(2), 635–636.

Bera, S., Misra, S., & Rodrigues, J. J. P. C. (2015). Cloud Computing Applications for Smart Grid: A Survey. *IEEE Transactions on Parallel and Distributed Systems, 26*(5), 1477–1494. doi:10.1109/TPDS.2014.2321378

Berger, L., Schwager, A., & Escudero-Garzás, J. (2013). Power Line Communications for Smart Grid Applications. *Journal of Electrical and Computer Engineering*. 10.1155/2013/712376

Bible, M., & Bivins, S. (2011). *Mastering Project Portfolio Management. A Systems approach to Achieving Strategic Objectives*. Fort Lauderdale, FL: J. Ross Publishing.

Biersack, A. (2006). Reimagining political ecology: culture/power/history/nature. In A. Biersack & J. B. Greenberg (Eds.), *Reimagining political ecology* (pp. 3–40). Durham, NC: Duke University Press. doi:10.1215/9780822388142-001

Birt, B. J., Newsham, G. R., Beausoleil-Morrison, I., Armstrong, M. M., Saldanha, N., & Rowlands, I. H. (2012). Disaggregating categories of electrical energy end-use from whole-house hourly data. *Energy and Buildings, 50*, 93–102. doi:10.1016/j.enbuild.2012.03.025

Blok, K. (2005). Improving Energy Efficiency by Five Percent and More per Year? *Journal of Industrial Ecology, 8*(4), 87–99. doi:10.1162/1088198043630478

Bocock, R. (1993). *Consumption.* London: Routledge. doi:10.4324/9780203313114

Borcoci, E. (2016). Fog-computing versus SDN/NFV and Cloud computing in 5G. Paper presented in DataSys conference, Valencia, Spain.

Boudreau, M. C., Chen, A., & Huber, M. (2008). Green IS: Building sustainable business practices. Information systems: A global text, 1-17.

Bouhafs, F., Mackay, M., & Merabti, M. (2012). Links to the future: Communication requirements and challenges in the smart grid. *IEEE Power & Energy Magazine, 10*(1), 24–32. doi:10.1109/MPE.2011.943134

Bryant, R. (1998). Power, knowledge and political ecology in the third world: A review. *Progress in Physical Geography, 22*(1), 79–94. doi:10.1177/030913339802200104

Brynjolfsson, E., & Mcelheran, K. (2016). The Rapid adoption of Data-Driven Decision Making. *The American Economic Review, 106*(5), 133–139. doi:10.1257/aer.p20161016

Buchmann, E., Boehm, K., Burghardt, T., & Kessler, S. (2013). Re-identification of Smart Meter data. *Personal and Ubiquitous Computing, 17*(4, SI), 653–662. 10.100700779-012-0513-6

Bumiller, G., Lampe, L., & Hrasnica, H. (2010). Power line communications for large-scale control and automation systems. *IEEE Communications Magazine, 48*(4), 106–113. doi:10.1109/MCOM.2010.5439083

Burmatova O.P. (2015a). Nature conservation strategy for regional socioeconomic development. *Region Research of Russia, 5*(3), 286-297.

Burmatova, O. P. (2015b). Environmental and Economic Diagnostics of the Local Production Systems. In Functioning of the local production systems in Bulgaria, Poland and Russia theoretical and economic policy issues. Lodz: Łódź University Press.

Byers, C. C., & Wetterwald, P. (2015). *Fog Computing Distributing Data and Intelligence for Resiliency and Scale Necessary for IoT the Internet of Things*. Paper presented at ACM Symposium on Ubiquity.

Candanedo, L. M., Feldheim, V., & Deramaix, D. (2017). Data driven prediction models of energy use of appliances in a low-energy house. *Energy and Buildings*, *140*, 81–97. doi:10.1016/j.enbuild.2017.01.083

Canvanizer 2.0. (2017). *Canvanizer*. Retrieved from https://canvanizer.com/new/open-innovation-canvas

Cao, H.-A., Wijaya, T. K., Aberer, K., & Nunes, N. (2017). Estimating human interactions with electrical appliances for activity-based energy savings recommendations. In *Proceedings - 2016 IEEE International Conference on Big Data, Big Data 2016* (pp. 1301–1308). IEEE. 10.1109/BigData.2016.7840734

Cardone, G., Foschini, L., Bellavista, P., Corradi, A., Borcea, C., Talasila, M., & Curtmola, R. (2013). Fostering participaction in smart cities: A geo-social crowdsensing platform. *IEEE Communications Magazine*, *51*(6), 112–119. doi:10.1109/MCOM.2013.6525603

Carli, R., Dotoli, M., Pellegrino, R., & Ranieri, L. (2013). Measuring and managing the smartness of cities: A framework for classifying performance indicators. In *Systems, Man, and Cybernetics (SMC), 2013 IEEE International Conference on* (pp. 1288-1293). IEEE.

Carnegie Mellon University. (n.d.). Retrieved from: https://www.cmu.edu

Carriero, D. (2015). United smart cities: Smart urban solutions for transition and developing countries. In *9th session of the WPLA*. UNECE. https://www.unece.org/fileadmin/DAM/hlm/wpla/sessions/9th_session/day_2_presentations/13_Carriero_WPLA9_day2.pdf

Castells, M. (2000). *Toward a Sociology of the Network Society*. American Sociological Association.

Cerci, N. (2015). Akıllı kentler bir ihtiyaç ve geleceğe dönük bir yaşam modelidir. *Bilişim Dergisi, 43*(172). Retrieved May 12, 2016, from http://www.bilisimdergisi. org/s172/FLASH/index.html

Cerrudo, C. (2015). *An emerging us (and world) threat: Cities wide open to cyber-attacks*. Securing Smart Cities.

Chakravorty, A., Rong, C., Evensen, P., & Wlodarczyk, T. W. (2014). A distributed gaussian-means clustering algorithm for forecasting domestic energy usage. *Proceedings of 2014 International Conference on Smart Computing, SMARTCOMP 2014*, 229–236. 10.1109/SMARTCOMP.2014.7043863

Chiky, R., Decreusefond, L., & Hébrail, G. (2010). Aggregation of asynchronous electric power consumption time series knowing the integral. In *Advances in Database Technology - EDBT 2010 - 13th International Conference on Extending Database Technology, Proceedings* (pp. 663–668). Academic Press. 10.1145/1739041.1739122

Cisco. (2016). *Fog Computing and the Internet of Things: Extend the Cloud to Where the Things Are*. Retrieved from http://www.cisco.com/c/dam/en_us/solutions/trends/iot/docs/computing-overview.pdf

Collier, S. E. (2010). Ten steps to a smarter grid. *IEEE Industry Applications Magazine, 16*(2), 62–68. doi:10.1109/MIAS.2009.935500

Conforto, E., Salum, F., Amaral, D., da Silva, S., & Magnanini de Almeida, L. (2014). Can Agile Project Management be Adopted by Industries Other than Software Development? *Project Management Journal, 45*(3), 21–34. doi:10.1002/pmj.21410

Conti, M., Das, S., Bisdikian, M., Kumar, M., Ni, L., Passarella, A., ... Zambonelli, F. (2012). Looking ahead in pervasive computing: Challenges and opportunities in the era of cyberphysical convergence. *Pervasive and Mobile Computing, 8*(1), 2–21. doi:10.1016/j.pmcj.2011.10.001

Crowley, J. (2003). *The invention of comfort*. Baltimore, MD: Johns Hopkins University Press.

Стратегия социально-экономического развития Сибири до 2020 года (Раздел IV. Приоритетные межотраслевые направления развития Сибири - Решение экологических проблем) – Утверждена распоряжением Правительства Российской Федерации от 5 июля 2010 г. N° 1120-р. (n.d.). [Strategy of socio-economic development of Siberia until 2020 (Section IV. Priority intersectoral directions of the development of Siberia -. Environmental issues) - Approved by the Federal Government on July 5, 2010 N° 1120-p]. Retrieved from http://www.sibfo.ru/strategia/strdoc.php

Dameri, R. P. (2013). Searching for smart city definition: A comprehensive proposal. *International Journal of Computers and Technology*, *11*(5), 2544–2551. doi:10.24297/ijct.v11i5.1142

Dastjerdi, A.V., & Buyya, R. (2017, January). *Fog Computing: Helping the Internet of Things Realize its Potential*. IEEE Computer Society.

Davenport, T., Barth, P., & Bean, R. (2012). How "Big Data" Is Different. *MIT Sloan Management Review*, 22–24.

Davidson, D., & Hatt, K. (2005). *Consuming Sustainability Critical Social Analysis of Ecological Change*. Fernwood Publishing.

De Jong, M., Joss, S., Schraven, D., Zhan, C., & Weijnen, M. (2015). Sustainable–smart–resilient–low carbon–eco–knowledge cities; making sense of a multitude of concepts promoting sustainable urbanization. *Journal of Cleaner Production*, *109*, 25–38. doi:10.1016/j.jclepro.2015.02.004

Department of Energy. (2010). Communications Requirements of Smart Grid Technologies. Washington, DC: Academic Press.

Dertouzos, M. (2001). Human-centered Systems. In P. Denning (Ed.), The Invisible Future (pp. 181-192). ACM Press.

Deshmukh, U. A., & More, S. A. (2016). Fog Computing: New Approach in the World of Cloud Computing. *International Journal of Innovative Research in Computer and Communication Engineering*, *4*(9), 16310–16316.

Digital Age. (2014). *Dünyadan Akıllı Şehir Örnekleri*. Retrieved April 07, 2017, from http://digitalage.com.tr/dunyadan-akilli-sehir-ornekleri/

Digital Agenda for Europe. (2014). *The EU explained: Digital agenda for Europe European Commission European Commission Directorate-General for Communication Citizens information 1049*. Publications Office of the European Union

Dobbyn, J., & Thomas, G. (2005). *Seeing the light: the impact of micro-generation on our use of energy*. Sustainable Consumption Roundtable, London, UK.

Dohler, M., Vilajosana, I., Vilajosana, X., & Llosa, J. (2011), December. Smart cities: An action plan. In *Proc. Barcelona Smart Cities Congress* (pp. 1-6). Academic Press.

Domingo, A., Bellalta, B., Palacin, M., Oliver, M., & Almirall, E. (2013). Public open sensor data: Revolutionizing smart cities. *IEEE Technology and Society Magazine*, *32*(4), 50–56. doi:10.1109/MTS.2013.2286421

Dong, M., Meira, P. C. M., Xu, W., & Chung, C. Y. (2013). Non-intrusive signature extraction for major residential loads. *IEEE Transactions on Smart Grid*, *4*(3), 1421–1430. doi:10.1109/TSG.2013.2245926

Dyson, M. E. H., Borgeson, S. D., Tabone, M. D., & Callaway, D. S. (2014). Using smart meter data to estimate demand response potential, with application to solar energy integration. *Energy Policy*, *73*, 607–619. doi:10.1016/j.enpol.2014.05.053

Eberle, U., Brohmann, B., & Graulich, K. (2004). Sustainable consumption needs visions. Position Paper. Institute of Applied Ecology, Öko-Institut, Freiburg/ Darmstadt.

El Sawy, O. (2003). The IS Core IX. The 3 Faces of IS Identity: Connection, Immersion, and Fusion. *Communications of the Association for Information Systems*, *12*, 588–598.

Emfinger, W., Dubey, A., Volgyesi, P., Sallai, J., & Karsai, G. (2016). Demo Abstract: RIAPS – A Resilient Information Architecture Platform for Edge Computing. Paper presented at IEEE/ACM Symposium on Edge Computing (SEC), Washington, DC. 10.1109/SEC.2016.23

ETH Zurich. (n.d.). *ETH Zurich*. Retrieved from: https://www.ethz.ch

European Commission. (2005). Doing More with Less: Green Paper on energy efficiency. Brussels: Author.

European Commission. (2010b). *Communication From the Commission to the European Parliament, The European Council, The Council.* Brussels: The European Central Bank, The Economic and Social Committee and the Committee of The Regions.

European Environment Agency (EEA). (2005). *Household consumption and the environment.* EEA Report No. 11/2005. Copenhagen: Author.

European Environment Agency. (2005). The European environment — State and outlook 2005. Copenhagen: Author.

Fang, X., Misra, S., Xue, G., & Yang, D. (2012). Smart grid—The new and improved power grid: A survey. *IEEE Communications Surveys and Tutorials, 14*(4), 944–980. doi:10.1109/SURV.2011.101911.00087

Fan, Z., Kulkarni, P., Gormus, S., Efthymiou, C., Kalogridis, G., Sooriyabandara, M., ... Chin, W. H. (2013). Smart Grid Communications: Overview of Research Challenges, Solutions, and Standardization Activities. *IEEE Communications Surveys and Tutorials, 15*(1), 21–38. doi:10.1109/SURV.2011.122211.00021

Farhangi, H. (2010). The Path of the Smart Grid 18. *IEEE Power & Energy Magazine, 8*, 18–28. doi:10.1109/MPE.2009.934876

Ferreira, H., Lampe, L., Newbury, J., & Swart, T. (2010). *Power Line Communications.* New York: Wiley. doi:10.1002/9780470661291

Fine, B., & Leopold, E. (1993). *The world of consumption.* London: Routledge.

Forfás. (2013). *Priority Area K Smart Grids and Smart Cities Action Plan July 2013.* Academic Press.

Foster, I., Kesselman, C., & Tuecke, S. (2001). The Anatomy of Grid: Enabling Scalable Virtual Organizations. *International Journal of Supercomputer Applications, 15*(3).

Freeman, R., & Reed, D. L. (1983). Stockholders and stakeholders: A new perspective on corporate governance. *California Management Review, 25*(3), 88–106. doi:10.2307/41165018

Fusco, F., Wurst, M., & Yoon, W. J. (2012). Mining Residential Household Information from Low-resolution Smart Meter Data. *21st International Conference on Pattern Recognition (ICPR 2012)*, 3545–3548.

Gajowniczek, K., & Zabkowski, T. (2015). Data mining techniques for detecting household characteristics based on smart meter data. *Energies*, *8*(7), 7407–7427. doi:10.3390/en8077407

Galli, S., Scaglione, A., & Wang, Z. (2011). For the grid and through the grid: The role of power line communications in the smart grid. *Proceedings of the IEEE*, *99*(6), 998–1027. doi:10.1109/JPROC.2011.2109670

Gama, F., Sjödin, D., & Frishammar, J. (2017). Managing interorganizational technology development: Project management practices for market- and science-based partnership. *Creativity and Innovation Management*, *26*(2), 115–127. doi:10.1111/caim.12207

Garcia-Hernandez, J. (2011). An analysis of communications and networking technologies for the smart grid. *Proceedings of CIGRE International Symposium, The Electric Power System of the Future*.

Gezer, C., & Buratti, C. (2011). A ZigBee smart energy implementation for energy efficient buildings. *Proceedings of 2011 IEEE 73rd Vehicular Technology Conference (VTC Spring)*, 1-5. 10.1109/VETECS.2011.5956726

Gharavi, H., & Ghafurian, R. (Eds.). (2011). Smart grid: The electric energy system of the future. Academic Press.

Giannuzzi, G. M., Pisani, C., Vaccaro, A., & Villacci, D. (2015). Overhead transmission lines dynamic line rating estimation in WAMS environments. *Proceedings of 2015 International Conference on Clean Electrical Power (ICCEP)*, 165-169. 10.1109/ICCEP.2015.7177618

Giordano, V., Gangale, F., Fulli, G., Jiménez, M. S., Onyeji, I., Colta, A., ... Maschio, I. (2011). *Smart Grid projects in Europe: lessons learned and current developments*. JRC Reference Reports, Publications Office of the European Union.

Gomes, D., Colunga, R., Gupta, P., & Balasubramanian, A. (2015). Distribution automation case study: Rapid fault detection, isolation, and power restoration for a reliable underground distribution system. *Proceedings of 2015 68th Annual Conference for Protective Relay Engineers*. 10.1109/CPRE.2015.7102176

Gontar, B., Gontar, Z., & Pamuła, A. (2013). Deployment of Smart City Concept in Poland. Selected Aspects *SIsteminiai Tyrimai, 67*(3), 39-51. 10.7220/MOSR.1392.1142.2013.67.3

Gubbi, J., Buyya, R., Marusic, S., & Palaniswami, M. (2013). Internet of Things (IoT): A vision, architectural elements, and future directions. *Future Generation Computer Systems, 29*(7), 1645–1660. doi:10.1016/j.future.2013.01.010

Gungor, V. C., Lu, B., & Hancke, G. P. (2010). Opportunities and challenges of wireless sensor networks in smart grid. *IEEE Transactions on Industrial Electronics, 57*(10), 3557–3564. doi:10.1109/TIE.2009.2039455

Gungor, V. C., Sahin, D., Kocak, T., Ergut, S., Buccella, C., Cecati, C., & Hancke, G. P. (2011). Smart grid technologies: Communication technologies and standards. *IEEE Transactions on Industrial Informatics, 7*(4), 529–539. doi:10.1109/TII.2011.2166794

Gungor, V. C., Sahin, D., Kocak, T., Ergut, S., Buccella, C., Cecati, C., & Hancke, G. P. (2013). A survey on smart grid potential applications and communication requirements. *IEEE Transactions on Industrial Informatics, 9*(1), 28–42. doi:10.1109/TII.2012.2218253

Guo, Y., Pan, M., & Fang, Y. (2012). Optimal power management of residential customers in the smart grid. *IEEE Transactions on Parallel and Distributed Systems, 23*(9), 1593–1606. doi:10.1109/TPDS.2012.25

Guo, Z., Wang, Z. J., & Kashani, A. (2015). Home Appliance Load Modeling From Aggregated Smart Meter Data. *IEEE Transactions on Power Systems, 30*(1), 254–262. doi:10.1109/TPWRS.2014.2327041

Gupta, H., Nath, S. B., Chakraborty, S., & Ghosh, S. K. (2016). *SDFog: A Software Defined Computing Architecture for QoS Aware Service Orchestration over Edge Devices*. arXiv preprint arXiv:1609.01190

Hall, P. (1988). *Cities of tomorrow*. Blackwell Publishers.

Hamilton, B., Miller, J., & Renz, B. (2010). *Understanding the benefits of smart grid*. Tech. Rep. DOE/NETL-2010/1413. U.S. Department of Energy.

Hancke, G. P., & Hancke, G. P. Jr. (2012). The role of advanced sensing in smart cities. *Sensors (Basel)*, *13*(1), 393–425. doi:10.3390130100393 PMID:23271603

Hannan, M., & Freeman, J. (1977). The population ecology of organizations. *American Journal of Sociology*, *82*(5), 929–964. doi:10.1086/226424

Harp, D. R., & Gregory-Brown, B. (2013). *IT/OT Convergence: Bridging the Divide*. NexDefense ICS SANS Report. Retrieved from https://ics.sans.org/media/IT-OT-Convergence-NexDefense-Whitepaper.pdf

Harrison, C., & Donnelly, I. A. (2011). A theory of smart cities. In *Proceedings of the 55th Annual Meeting of the ISSS-2011* (*Vol. 55*, No. 1). Academic Press.

Harrison, C., Eckman, B., Hamilton, R., Hartswick, P., Kalagnanam, J., Paraszczak, J., & Williams, P. (2010). Foundations for smarter cities. *IBM Journal of Research and Development*, *54*(4), 1–16. doi:10.1147/JRD.2010.2048257

Hart, D. G. (2008). Using AMI to realize the Smart Grid. *Proceedings of 2008 IEEE Power and Energy Society General Meeting - Conversion and Delivery of Electrical Energy in the 21st Century*. 10.1109/PES.2008.4596961

Harvey, D. (2005). *Brief History of Neoliberalism*. New York: Oxford University Press.

Hayes, B., Gruber, J., & Prodanovic, M. (2015). Short-Term Load Forecasting at the local level using smart meter data. *IEEE Eindhoven PowerTech*, 1–6. 10.1109/PTC.2015.7232358

Henderson, J., & Venkatraman, N. (1993). Strategic Alignment: Levering Information Technology for Transforming Organizations. *IBM Systems Journal*, *32*(1), 4–16. doi:10.1147j.382.0472

Hernández, L., Baladrón, C., Aguiar, J. M., Calavia, L., Carro, B., Sánchez-Esguevillas, A., ... Gómez, J. (2012). A study of the relationship between weather variables and electric power demand inside a smart grid/smart world framework. *Sensors (Basel)*, *12*(9), 11571–11591. doi:10.3390120911571

Herzog, A. V., Lipman, T. E., & Kammen, D. M. (2001). Renewable energy sources. In Encyclopedia of Life Support Systems (EOLSS). Forerunner Volume- 'Perspectives and Overview of Life Support Systems and Sustainable Development. Academic Press.

Hiller, J. S., & Blanke, J. M. (2016). *Smart Cities*. Big Data, and the Resilience of Privacy.

Hollands R. G. (2008). Will the Real Smart City Please Stand Up? *City*, *12*(3), 303-320.

Hughes, T. (1983). *Networks of power: Electrification in Western society*. Baltimore, MD: Johns Hopkins University Press.

Iansiti, M., & Levien, R. (2004). *The keystone advantage: what the new dynamics of business ecosytems mean for strategy, innovation and sustainability*. Boston: Harvard BusinessSchool Press.

Iansiti, M., & Levien, R. (2004a). Strategy as ecology. *Harvard Business Review*, *82*, 68–78. PMID:15029791

IEA. (2009b). *Comparative Study on Rural Electrification Policies in Emerging Countries*. Paris.Available at http://www.iea.org/papers/2010/rural_elect.Accessed 18/7/2017

Ishii, H., Kimino, K., Aljehani, M., Ohe, N., & Inoue, M. (2016). An Early Detection System for Dementia using the M2M/IoT Platform. *Procedia Computer Science*, *96*, 1332–1340. doi:10.1016/j.procs.2016.08.178

Jain, C. (2015). *UTC India-Student Essay-Inclusive Smart City*. Retrieved March 02, 2017, from http://utc.niua.org/utc-india-student-essay-inclusive-smart-city/

Jin, N., Flach, P., Wilcox, T., Sellman, R., Thumim, J., & Knobbe, A. (2014). Subgroup discovery in smart electricity meter data. *IEEE Transactions on Industrial Informatics*, *10*(2), 1327–1336. doi:10.1109/TII.2014.2311968

Johnson, P. (2009). *Human centered information integration for the smart grid.* Technical Report CSDL-09-15, University of Hawaii, Honolulu, HI.

Jugdev, K., Thomas, J., & Delisle, C. (2001). Rethinking project management old truths and new insights. *Project Management*, 36-43.

Kalkınma Bakanlığı, T. C. (2013). *Bilgi toplumu stratejisinin yenilenmesi projesi bilgi ve iletişim teknolojileri destekli yenilikçi çözümler ekseni küresel eğilimler ve ülke incelemeleri raporu.* Retrieved March 24, 2017, from www.bilgitoplumustratejisi.org

Kane, G., Palmer, D., Phillips, A., Kiron, D., & Buckley, N. (2015). *Strategy, not technology, drives digital transformation.* MIT Sloan Management Review.

Kannberg, L., Kintner-Meyer, C., Chassin, D., Pratt, R., DeSteese, J., Schienbein, L., . . . Warwick, W. (2003). *Grid Wise: The Benefits of a Transformed Energy System.* Pacific Northwest National Laboratory under contract with the United States Department of Energy: 25. arXiv:nlin/0409035.

Karhu, K., Botero, A., Vihavainen, S., Tang, T., & Hämäläinen, M. (2011). A Digital Ecosystem for Co-Creating Business with People. *Journal of Emerging Technologies in Web Intelligence*, *3*(3), 197–205. doi:10.4304/jetwi.3.3.197-205

Kavousian, A., Rajagopal, R., & Fischer, M. (2013). Determinants of residential electricity consumption: Using smart meter data to examine the effect of climate, building characteristics, appliance stock, and occupants' behavior. *Energy*, *55*, 184–194. doi:10.1016/j.energy.2013.03.086

Kennedy, C., Steinberger, J., Gasson, B., Hansen, Y., Hillman, T., Havranek, M., . . . Mendez, G. V., (2009). *Greenhouse gas emissions from global cities.* Academic Press.

Khansari, N., Mostashari, A., & Mansouri, M. (2014). Impacting sustainable behavior and planning in smart city. *International Journal of Sustainable Land Use and Urban Planning*, *1*(2). doi:10.24102/ijslup.v1i2.365

Khedkar, S. V., & Gawande, A. D. (2014). Data Partitioning Technique to Improve Cloud Data Storage Security. *International Journal of Computer Science and Information Technologies*, 5(3), 3347–3350.

Kim, J., Filali, F., & Ko, Y. B. (2015). Trends and potentials of the smart grid infrastructure: From ICT, sub-system to SDN-enabled smart grid architecture. *Applied Sciences*, 5(4), 706–727. doi:10.3390/app5040706

KIT. (n.d.). *Industry 4.0 Collaboration Lab*. Karlsruher Institut for Technologie. Retrieved from: https://www.imi.kit.edu/2754.php

Kline, W., Hixson, C., Mason, T., Brackin, P., Bunch, R., & Dee, K. (2013). The Innovation Canvas - A Tool to Develop Integrated Product Designs and Business Models. In *120th ASEE Annual Conference and Exposition* (pp. 1-11). Atlanta, GA: American Society for Engineering Education.

Klotz, F. (2016, August). Navigating the Leadership Challenges of Innovation Ecosystems. *MIT SMR Frontiers*, 1-5.

Kogias, D., Tuna, G., & Gungor, V. C. (2016). Cognitive Radio. In H. T. Mouftah & M. Erol-Kantarci (Eds.), *Smart Grid: Networking, Data Management, and Business Models*. Boca Raton, FL: CRC Press.

Kouzelis, K., & Bak-jensen, B. (2014). *A Simplified Short Term Load Forecasting Method Based on Sequential Patterns*. Academic Press.

Kranz, J., Kolbe, L. M., Koo, C., & Boudreau, M. C. (2015). Smart energy: Where do we stand and where should we go? *Electronic Markets*, 25(1), 7–16. doi:10.100712525-015-0180-3

Kuleshov, V.V., & Seliverstov, V.E. (2016). Program for reindustrialization of the Novosibirsk Oblast Economy: Development ideology and main directions of implementation. *Regional Research of Russia*, 6(3), 214-226.

Kumar, P. (2015). *What's the Real Mean of "Smart City"?* Retrieved March 12, 2017, from http://www.smartcitiesprojects.com/whats-the-real-mean-of-smart-city/

Kuyper, T.S.T. (2016). *Smart City Strategy & Upscaling: Comparing Barcelona and Amsterdam*. Academic Press.

Kuzlu, M., Pipattanasomporn, M., & Rahman, S. (2014). Communication network requirements for major smart grid applications in HAN, NAN and WAN. *Computer Networks*, *67*, 74–88. doi:10.1016/j.comnet.2014.03.029

Kwac, J., Flora, J., & Rajagopal, R. (2014). Household Energy Consumption Segmentation Using Hourly Data. *IEEE Transactions on Smart Grid*, *5*(1), 420–430. doi:10.1109/TSG.2013.2278477

Lavin, A., & Klabjan, D. (2015). Clustering time-series energy data from smart meters. *Energy Efficiency*, *8*(4), 681–689. doi:10.100712053-014-9316-0

Lehmann, S. (2010). Green Urbanism. *Formulating a Series of Holistic Principles Surveys and Perspectives Integrating Environment and Society*, *3*(2), 1–10.

Liao, Y.-S., Liao, H.-Y., Liu, D.-R., Fan, W.-T., & Omar, H. (2016). Intelligent Power Resource Allocation by Context-Based Usage Mining. In *Proceedings - 2015 IIAI 4th International Congress on Advanced Applied Informatics, IIAI-AAI 2015* (pp. 546–550). IIAI. 10.1109/IIAI-AAI.2015.165

Lichtenberg, S. (2010). *Smart Grid Data: Must There Be Conflict Between Energy Management and Consumer Privacy?* National Regulatory Research Institute.

Li, H., Gong, S., Lai, L., Han, Z., Qiu, R. C., & Yang, D. (2012). Efficient and secure wireless communications for advanced metering infrastructure in smart grids. *IEEE Transactions on Smart Grid*, *3*(3), 1540–1551. doi:10.1109/TSG.2012.2203156

Ling, A. P. A., Kokichi, S., & Masao, M. (2012). *The Japanese smart grid initiatives, investments, and collaborations*. arXiv preprint arXiv:1208.5394

Liserre, M., Sauter, T., & Hung, J. Y. (2010). Future energy systems: Integrating renewable energy sources into the smart power grid through industrial electronics. *IEEE Industrial Electronics Magazine*, *4*(1), 18–37. doi:10.1109/MIE.2010.935861

Li, Z., Wang, P., Chu, Z., Zhu, H., Sun, Z., & Li, Y. (2013). A three-phase 10 kVAC-750 VDC power electronic transformer for smart distribution grid. *Proceedings of 2013 15th European Conference on Power Electronics and Applications (EPE)*. 10.1109/EPE.2013.6631810

Lloyd-Walker, B., Mills, A., & Walker, D. (2014). Enabling construction innovation: The roles of a no-blame culture as a collaboration behavioural driver in project alliances. *Construction Management and Economics*, *32*(3), 229–245. doi:10.108 0/01446193.2014.892629

Loebbecke, C., & Picot, A. (2015). Reflections on societal and business model transformation arising from digitization and big data analytics: A research agenda. *The Journal of Strategic Information Systems*, *24*(3), 149–157. doi:10.1016/j. jsis.2015.08.002

Lombardi, P. (2011). New challenges in the evaluation of Smart Cities. *Network Industries Quarterly*, *13*(3), 8–10.

Lu, D., Kanchev, H., Colas, F., Lazarov, V., & Francois, B. (2011). Energy management and operational planning of a microgrid with a PV-based active generator for smart grid applications. *IEEE Transactions on Industrial Electronics*, *58*(10), 4583–4592. doi:10.1109/TIE.2011.2119451

Lugaric, L., Krajcar, S., & Simic, Z. (2010). Smart City - Platform for Emergent Phenomena Power System Testbed Simulator. *Innovative Smart Grid Technologies Conference*. IEEE Xplore. Retrieved March 30, 2017, http://ieeexplore.ieee.org/stamp/stamp.jsp?arnumber=5638890

Luntovskyy, A., & Spillner, J. (2017). *Smart Grid, Internet of Things and Fog Computing. Architectural Transformations in Network Services and Distributed Systems*. Wiesbaden, Germany: Springer. doi:10.1007/978-3-658-14842-3

Madsen, H., Burtschy, B., Albeanu, G., & Popentiu-Vladicescu, F. (2013). *Reliability in the utility computing era: Towards reliable Fog Computing*. Paper presented at 20th International Conference on Systems, Signals and Image Processing, Bucharest, Romania. 10.1109/IWSSIP.2013.6623445

Madureira, A., Gouveia, C., Moreira, C., Seca, L., & Lopes, J. P. (2013). Coordinated management of distributed energy resources in electrical distribution systems. *2013 IEEE PES Conference on Innovative Smart Grid Technologies, ISGT LA 2013*. 10.1109/ISGT-LA.2013.6554446

Mahmood, A., Aamir, M., & Anis, M. I. (2008). Design and implementation of AMR Smart Grid System. *Proceedings of 2008 IEEE Canada Electric Power Conference (EPEC 2008).* 10.1109/EPC.2008.4763340

Majava, J., Harkonen, J., & Haapasalo, H. (2015). The relations between stakeholders and product development drivers: Practitioners' perspectives. *International Journal of Innovation and Learning, 17*(1), 59–78. doi:10.1504/IJIL.2015.066064

Marinescu, A., Harris, C., Dusparic, I., Clarke, S., & Cahill, V. (2013). Residential electrical demand forecasting in very small scale: An evaluation of forecasting methods. *2013 2nd International Workshop on Software Engineering Challenges for the Smart Grid, SE4SG 2013 - Proceedings*, 25–32. 10.1109/SE4SG.2013.6596108

Markus, L., & Loebbecke, C. (2013). Commositized digital processes and business community platforms: New opportunities and challenges for digital business strategies. *Management Information Systems Quarterly, 37*(2), 649–653.

Matsumoto, S., Serizawa, Y., Fujikawa, F., Shioyama, T., Ishihara, Y., Katayama, S., ... Ishibashi, A. (2012). Wide-Area Situational Awareness (WASA) system based upon international standards. *Proceedings of 11th International Conference on Developments in Power Systems Protection (DPSP 2012).* 10.1049/cp.2012.0032

Matugina, E. G., Egorova, A. Y., & Palamarchuk, A. V. (2015, May). To the question of forming and development of ecological entrepreneurship. *SWorld Journal. Economy*, 66-68.

Mazza, P. (2004). The Smart Energy Network: Electricity's Third Great Revolution. *Climate Solutions, 1*(2), 1–7.

McAfee, A., & Brynjolfsson, E. (2012). Big Data: The Management Revolution. *Harvard Business Review, 10*, 60–79. PMID:23074865

McDaniel, P., & McLaughlin, S. (2009). Security and privacy challenges in the smart grid. *IEEE Security and Privacy, 7*(3), 75–77. doi:10.1109/MSP.2009.76

McKinsey and Company (2010). *A lean approach to energy efficiency.* Author.

McKinsey. (n.d.). *McKinsey & Company.* Retrieved from: https://www.mckinsey.com/

McLoughlin, F., Duffy, A., & Conlon, M. (2015). A clustering approach to domestic electricity load profile characterisation using smart metering data. *Applied Energy*, *141*, 190–199. doi:10.1016/j.apenergy.2014.12.039

Meiling, S., Schmidt, T. C., & Steinbach, T. (2015). On performance and robustness of internet-based smart grid communication: A case study for Germany. *Proceedings of 2015 IEEE International Conference on Smart Grid Communications (SmartGridComm)*. 10.1109/SmartGridComm.2015.7436316

Meliopoulos, A. P. S., Polymeneas, E., Tan, Z., Huang, R., & Zhao, D. (2013). Advanced Distribution Management System. *IEEE Transactions on Smart Grid*, *4*(4), 2109–2117. doi:10.1109/TSG.2013.2261564

Metke, A. R., & Ekl, R. L. (2010). Security technology for smart grid networks. *IEEE Transactions on Smart Grid*, *1*(1), 99–107. doi:10.1109/TSG.2010.2046347

Mohd, A., Ortjohann, E., Schmelter, A., Hamsic, N., & Morton, D. (2008). Challenges in integrating distributed energy storage systems into future smart grid. In *Industrial Electronics, 2008. ISIE 2008. IEEE International Symposium on* (pp. 1627-1632). IEEE.

Momoh, J. A. (2009). Smart grid design for efficient and flexible power networks operation and control. In Power Systems Conference and Exposition, 2009. PSCE'09. IEEE/PES (pp. 1-8). IEEE.

Morvaj, B., Lugaric, L., & Krajcar, S. (2011). Demonstrating smart buildings and smart grid features in a smart energy city. In *Energetics (IYCE), Proceedings of the 2011 3rd International Youth Conference on* (pp. 1-8). IEEE.

Moslehi, K., & Kumar, R. (2010). A reliability perspective of the smart grid. *IEEE Transactions on Smart Grid*, *1*(1), 57–64. doi:10.1109/TSG.2010.2046346

Mulgan, G. (1991). *Communication and Control*. Cambridge, UK: Polity Press.

Mwasilu, F., Justo, J. J., Kim, E. K., Do, T. D., & Jung, J. W. (2014). Electric vehicles and smart grid interaction: A review on vehicle to grid and renewable energy sources integration. *Renewable & Sustainable Energy Reviews*, *34*, 501–516. doi:10.1016/j.rser.2014.03.031

National Electric Reliability Commission. (2012). *CIP-004-6: Cyber Security – Personnel & Training*. Retrieved from http://www.nerc.com/pa/Stand/Prjct2014XXCrtclInfraPrtctnVr5Rvns/CIP-004-6_CLEAN_06022014.pdf

National Electric Reliability Commission. (2014). *CIP-008-5: Cyber Security – Incident Reporting and Response Planning*. Retrieved from: http://www.nerc.com/pa/Stand/Project%20200806%20Cyber%20Security%20Order%20706%20DL/CIP-008-5_clean_4_(2012-1024-1218).pdf

National Electric Reliability Commission. (2014). *CIP-010-2: Cyber Security – Configuration Change Management and Vulnerability Assessments*. Retrieved from http://www.nerc.com/pa/Stand/Reliability%20Standards/CIP-010-2.pdf

National Electric Reliability Commission. (2014). *CIP-011-2: Cyber Security – Information Protection*. Retrieved from http://www.nerc.com/pa/Stand/Prjct2014XXCrtclInfraPrtctnVr5Rvns/CIP-011-2_CLEAN_06022014.pdf

Nelson, N., (2016). *The Impact of Dragonfly Malware on Industrial Control Systems*. SANS Institute InfoSec Reading Room Report.

Nijkamp, P., & Pepping, G. (1998). A Meta-analytical evaluation of sustainable city initiatives. *Urban Studies (Edinburgh, Scotland), 35*(9), 1481–1500. doi:10.1080/0042098984240

Niska, H. (2015). *Evolving Smart Meter Data Driven Model for Short- Term Forecasting of Electric Loads*. Academic Press.

Nuottila, J., Aaltonen, K., & Kujala, J. (2016). Challenges of adopting agile methods in a public organization. *International Journal of Information Systems and Project Management, 4*(3), 65–85.

Oestreicher-Singer, G., & Zalmanson, L. (2013). Content or Community? A Digital Business Strategy for Content Providers in the Social Age. *Management Information Systems Quarterly, 37*(2), 591–616. doi:10.25300/MISQ/2013/37.2.12

Okay, F. Y., & Ozdemir, S. (2016) A Fog Computing Based Smart Grid Model. In *Proceedings of the International Symposium on Networks, Computers and Communications (ISNCC'16)*. IEEE. 10.1109/ISNCC.2016.7746062

Olson, N., Nolin, J., & Nelhans, G. (2015). Semantic web, ubiquitous computing, or internet of things? A macro-analysis of scholarly publications. *The Journal of Documentation, 71*(5), 884–916. doi:10.1108/JD-03-2013-0033

OpenFog Consortium Architecture Working Group. (2016). *OpenFog Architecture Overview*. Retrieved from https://www.openfogconsortium.org/wp-content/uploads/OpenFog-Architecture-Overview-WP-2-2016.pdf

OpenFog Consortium Architecture Working Group. (2017). *OpenFog Reference Architecture for Fog Computing*. Retrieved from https://www.openfogconsortium.org/wp-content/uploads/OpenFog_Reference_Architecture_2_09_17-FINAL.pdf

Orwat, C., Graefe, A., & Faulwasser, T. (2008). Towards pervasive computing in health care - A literature review. *BMC Medical Informatics and Decision Making, 8*(26). PMID:18565221

Osterwalder, A., & Pigneur, Y. (2010). *Business Model Generation: A Handbook for Visionaries, Game Changers, and Challengers*. Hoboken, NJ: John Wiley & Sons, Inc.

Pacific Northwest National Laboratory. (2007). *Gridwise History: How did GridWise start?* Available at http:/gridwise.pnl.gov/foundations/history.stmAccessed 20/9/2017

Palensky, P., & Dietrich, D. (2011). Demand side management: Demand response, intelligent energy systems, and smart loads. *IEEE Transactions on Industrial Informatics, 7*(3), 381–388. doi:10.1109/TII.2011.2158841

Parker, G., Van Alstyne, M., & Choudary, S. (2016). *Platform Revolution. How networked markets are transforming the economy and how to make them work for you*. New York: W. W. Norton & Company.

Park, S., Ryu, S., Choi, Y., Kim, J., & Kim, H. (2015). Data-driven baseline estimation of residential buildings for demand response. *Energies, 8*(9), 10239–10259. doi:10.3390/en80910239

Parson, O., Ghosh, S., Weal, M., & Rogers, A. (2014). An unsupervised training method for non-intrusive appliance load monitoring. *Artificial Intelligence, 217*, 1–19. doi:10.1016/j.artint.2014.07.010

Pirisi, A., Grimaccia, F., Mussetta, M., & Zich, R. E. (2012). Novel speed bumps design and optimization for vehicles' energy recovery in smart cities. *Energies*, *5*(11), 4624–4642. doi:10.3390/en5114624

Popeanga, J. (2012). Cloud computing and smart grids. *Database Systems Journal*, *3*(3), 57–66.

Porter, M., & Heppelmann, J. (2014). How Smart, Connected Products Are Transforming Competition. *Harvard Business Review*, 4–23.

Porter, M., & Heppelmann, J. (2015). How Smart, Connected Products are Transforming Companies. *Harvard Business Review*, 1–19.

Pradhan, S., Dubey, A., Khare, S., Sun, F., Sallai, J., Gokhale, A., . . . Sturm, M. (2016). *Poster Abstract: A Distributed and Resilient Platform for City-Scale Smart Systems*. Paper presented at IEEE/ACM Symposium on Edge Computing (SEC), Washington, DC. 10.1109/SEC.2016.28

Project Management Institute, Inc. (2008). *A Guide to the Project Management body of Knowledge (PMBOK Guide)-Forth Edition*. Newtown Square, PA: PMI Publications.

Quilumba, F. L., Lee, W. J., Huang, H., Wang, D. Y., & Szabados, R. (2014). An overview of AMI data preprocessing to enhance the performance of load forecasting. *2014 IEEE Industry Application Society Annual Meeting, IAS 2014*, 1–7. 10.1109/IAS.2014.6978369

Quilumba, F. L., Lee, W.-J., Huang, H., Wang, D. Y., & Szabados, R. L. (2015). Using Smart Meter Data to Improve the Accuracy of Intraday Load Forecasting Considering Customer Behavior Similarities. *IEEE Transactions on Smart Grid*, *6*(2), 911–918. doi:10.1109/TSG.2014.2364233

Rao, T.V.N., & Khan, M.A., Maschendra, M., & Kumar, M.K. (2015). A Paradigm Shift from Cloud to Fog Computing. *International Journal of Computer Science & Engineering Technology*, *5*(11), 385–389.

Rawat, D. B., & Bajracharya, C. (2015). Cyber security for smart grid systems: Status, challenges and perspectives. *Proceedings of SoutheastCon 2015*, 1-6. 10.1109/SECON.2015.7132891

Register, R. (1987). *Ecocity Berkeley: Building Cities for a Healthy Future.* North Atlantic Books.

Reim, W., Parida, V., & Örtqvist, D. (2015). Product-Service Systems (PSS) business models and tactics - a systematic literature review. *Journal of Cleaner Production, 97*, 61–75. doi:10.1016/j.jclepro.2014.07.003

Rennings, K., Brohmann, B., Nentwich, J., Schleich, J., Traber, T., & Wüstenhagen, R. (Eds.). (2013). *Sustainable Energy Consumption in Residential Buildings.* Springer. doi:10.1007/978-3-7908-2849-8

Resolution of the UN General Assembly adopted on 25 September 2015. (n.d.). Retrieved from https://documents-dds-ny.un.org/doc/UNDOC/GEN/N15/291/92/PDF/N1529192.pdf?OpenElement

Rodríguez-Molina, J., Martínez-Núñez, M., Martínez, J. F., & Pérez-Aguiar, W. (2014). Business models in the smart grid: Challenges, opportunities and proposals for prosumer profitability. *Energies, 7*(9), 6142–6171. doi:10.3390/en7096142

Roncero, J. R. (2008). Integration is key to smart grid management. In *Smart Grids for Distribution, 2008. IET-CIRED. CIRED Seminar* (pp. 1-4). IET.

Ronfeldt, D. (1992). Cyberocracy is coming. *The Information Society, 8*(4), 243–296. doi:10.1080/01972243.1992.9960123

Ross, J., Sebastian, I., & Beath, C. (2016). How to Develop a Great Digital Strategy. *MIT Sloan Management Review.*

Saber, A. Y., & Venayagamoorthy, G. K. (2011). Plug-in vehicles and renewable energy sources for cost and emission reductions. *IEEE Transactions on Industrial Electronics, 58*(4), 1229–1238. doi:10.1109/TIE.2010.2047828

Saharan, K. P., & Kumar, A. (2015). Fog in Comparison to Cloud: A Survey. *International Journal of Computers and Applications, 122*(3), 10–12. doi:10.5120/21679-4773

Sajjad, H. P., Danniswara, K., Al-Shishtawy, A., & Vlassov, V. (2016). *SpanEdge: Towards Unifying Stream Processing over Central and Near-the-Edge Data Centers.* Paper presented at IEEE/ACM Symposium on Edge Computing (SEC), Washington, DC. 10.1109/SEC.2016.17

Saleh, M., Althaibani, A., Esa, Y., Mhandi, Y., & Mohamed, A. (2015). Impact of clustering microgrids on their stability and resilience during blackouts. *2015 International Conference on Smart Grid and Clean Energy Technologies (ICSGCE)*, 195–200. 10.1109/ICSGCE.2015.7454295

Sassen,S .(2005). The Global City: introducing a Concept. *The Brown Journal of World Affairs, 11*(2), 27-43.

Sauter, T., & Lobashov, M. (2011). End-to-End communication architecture for smart grids. *IEEE Transactions on Industrial Electronics, 58*(4), 1218–1228. doi:10.1109/TIE.2010.2070771

Savio, D., Karlik, L., & Karnouskos, S. (2010). Predicting energy measurements of service-enabled devices in the future smartgrid. *UKSim2010 - UKSim 12th International Conference on Computer Modelling and Simulation*, 450–455. 10.1109/UKSIM.2010.89

Saxena, N., & Choi, B. J. (2015). State of the art authentication, access control, and secure integration in smart grid. *Energies, 8*(10), 11883–11915. doi:10.3390/en81011883

Schor, J. (2001). *Why Do We Consume So Much?* Clemens Lecture Series 13 Saint John's University.

Schor, J. (2005). Prices and quantities: Unsustainable consumption and the global economy. *Ecological Economics, 55*(3), 309–320. doi:10.1016/j.ecolecon.2005.07.030

Schuurman, D., Baccarne, B., De Marez, L., & Mechant, P. (2012). Smart ideas for smart cities: Investigating crowdsourcing for generating and selecting ideas for ICT innovation in a city context. *Journal of Theoretical and Applied Electronic Commerce Research, 7*(3), 49–62. doi:10.4067/S0718-18762012000300006

Seliverstov, V.E. (2012). Regional monitoring: Information management framework for regional policy and strategic planning. *Regional Research of Russia, 2*(1), 60-73.

Seliverstov, V.E., & Melnikova, L.V. (2013). Analysis of Strategic Planning in Regions of the Siberian Federal District. *Regional Research of Russia, 3*(1), 96-102.

Shenhar, A., & Dvir, D. (2007). *Reinventing project management: The diamond approach to successful growth and innovation.* Boston: Harvard Business Press.

Shenhar, A., Holzman, V., Melaned, B., & Zhao, Y. (2016). The Challenge of Innovation in Highly Complex Projects: What Can We Learn from Boeing's Dreamliner Experience? *Project Management Journal, 47*(2), 62–78. doi:10.1002/pmj.21579

Shirky, C. (2008). *Here Comes Everybody: How Change Happens when People Come together*. London: Penguin Books.

Shi, W., & Dustdar, S. (2016, May 13). The Promise of Edge Computing. *IEEE Computer Society, 49*(5), 78–81. doi:10.1109/MC.2016.145

Shove, E. (2003). Converging conventions of comfort, cleanliness and convenience. *Journal of Consumer Policy, 26*(4), 395–418. doi:10.1023/A:1026362829781

Sınmaz, S. (2013). The concept of "smart settlement" and basic principles in the framework of new developing planning approaches. *Megaron, 8*(2), 76–86. doi:10.5505/megaron.2013.35220

Slama, D., Puhlmann, F., Morrish, J., & Bhatnagar, R. (2015). *Enterprise IoT*. O'Reilly Media, Inc.

Smart Cities Council. (n.d.). Retrieved March 28, 2017, from http://smartcitiescouncil.com/

Smart Cities in Europe. (2016). Retrieved May 14, 2016, from, ftp://zappa.ubvu.vu.nl/20090048.pdf

Smart Cities Mission. (2016). Retrieved May 30, 2016, from http://smartcities.gov.in/writereaddata/SmartCityGuidelines.pdf

Smart Cities Readiness Guide. (2015). *The planning manual for building tomorrow's cities today*. Smart Cities Council. Retrieved February 12, 2017, from http://www.uraia.org/OlalaCMS4/files/574_arquivoB.pdf

Songdo, I. B. D. (2015). Retrieved May 14, 2016, from http://www.songdo.com

Sornalakshmi, K., & Vadivu, G. (2015). A Survey on Realtime Analytics Framework for Smart Grid Energy Management. *International Journal of Innovative Research in Science, Engineering and Technology, 4*(3), 1054–1058.

Spaargaren, G. (1997). *The ecological modernisation of production and consumption: in environmental. sociology*. Wageningen, The Netherlands: Landbouw University Wageningen.

Speake, S. (2005). Water:A Human Right. In *Consuming Sustainability Critical Social Analysis of Ecological Change*. Fernwood Publishing.

Stacey, R. (1995). The science of compelxity: An alternative perspective for strategic change processes. *Strategic Management Journal*, *16*(6), 477–495. doi:10.1002mj.4250160606

Stankovic, L., Stankovic, V., Liao, J., & Wilson, C. (2016). Measuring the energy intensity of domestic activities from smart meter data. *Applied Energy*, *183*, 1565–1580. doi:10.1016/j.apenergy.2016.09.087

Steinert, K., Marom, R., Richard, P., & Veiga, G., & Witters, L. (2011). Making cities smart and sustainable. *The Global Innovation Index*, *2011*, 87–95.

Sting, F., Loch, C., & Stempfhuber, D. (2015). Accelerating Projects by Encouraging Help. *MIT Sloan Management Review*, 5–13.

Strickland, E. (2011). Cisco bets on South Korean smart city. *IEEE Spectrum*, *48*(8), 11–12. doi:10.1109/MSPEC.2011.5960147

Stroustrup. (n.d.). Bjarne Stroustrup's FAQ. *Stroustrup*. Retrieved from: http://www.stroustrup.com/bs_faq.html#really-say-that

Succar, S., & Cavanagh, R. (2012). *The Promise of the Smart Grid: Goals, Policies, and Measurement Must Support Sustainability Benefits*. NRDC Issue Brief.

Su, W., Eichi, H., Zeng, W., & Chow, M. Y. (2012). A survey on the electrification of transportation in a smart grid environment. *IEEE Transactions on Industrial Informatics*, *8*(1), 1–10. doi:10.1109/TII.2011.2172454

Svejvig, P., & Andersen, P. (2015). Rethinking project management: A structured literature review with a critical look at the brave new world. *International Journal of Project Management*, *33*(2), 278–290. doi:10.1016/j.ijproman.2014.06.004

Tanoto, Y., & Setiabudi, D. (2016). Development of autonomous demand response system for electric load management. In *2016 Asian Conference on Energy, Power and Transportation Electrification, ACEPT 2016* (pp. 1-6). Marina Bay Sands: IEEE. 10.1109/ACEPT.2016.7811532

Tawde, R., Nivangune, A., & Sankhe, M. (2015). Cyber security in smart grid SCADA automation systems. *Proceedings of 2015 International Conference on Innovations in Information, Embedded and Communication Systems (ICIIECS)*, 1-5. 10.1109/ICIIECS.2015.7192918

Teece, D. (2007). Explicating dynamic capabilities: The nature and microfoundations of (sustainable) enterprise performance. *Strategic Management Journal, 28*(13), 1319–1350. doi:10.1002mj.640

Thompson, C. (1996). Caring consumers: Gendered consumption meanings and the juggling lifestyle. *The Journal of Consumer Research, 22*(4), 388–407. doi:10.1086/209457

Toffler, A. (1980). *The Third Wave*. Bantam Books.

Toffler, A., & Toffler, H. (1995). *Creating A New Civilization: The Politics Of The Third Wave*. Turner Publishing.

Tuna, G., Gungor, V. C., & Gulez, K. (2013). Wireless sensor networks for smart grid applications: A case study on link reliability and node lifetime evaluations in power distribution systems. *International Journal of Distributed Sensor Networks, 2013*, 1–11.

U.S. Department of Energy. (2003). *"Grid 2030" A National Vision for Electricity's Second 100 Years*. Office of Electric Transmission and Distribution.

U.S. Department of Energy. (2004). *National Electric Delivery Technologies Roadmap*. Office of Electric Transmission and Distribution.

UDHB. (2014). *Ulusal akıllı ulaşım sistemleri strateji belgesi ve eylem planı*. Ankara: T.C. Ulaştırma, Denizcilik ve Haberleşme Bakanlığı.

Ugale, B., Soni, P., Pema, T., & Patil, A. (2011). Role of cloud computing for smart grid of India and its cyber security. *Proc. IEEE Nirma Univ. Int. Conf. Eng.*, 1–5. 10.1109/NUiConE.2011.6153298

Ulusal Ulastiurma Portali. (n.d.). Retrieved March 28, 2017, from www.ulasim.gov.tr

UN. (1987). *Our common future report*. UN Documents. http://www.un-documents.net/wced-ocf.htm

UN. (2015). *World population prospects the 2015 revision*. New York: Department of Economic and Social Affairs.

UN. (2016). *World urbanization prospects*. Department of Economic and Social Affairs, Population Division.

UNEP/Wuppertal Institute Collaborating Centre on Sustainable Consumption and Production (CSCP). (2005) *Sustainable Energy Consumption*. A Background Paper prepared for the European Conference under the Marrakech Process on Sustainable Consumption and Production (SCP), Berlin, Germany.

UNESCO-MOST. (1996). Conference report on sustainability as a social science concept. Frankfurt: Author.

Uribe-Pérez, N., Hernández, L., de la Vega, D., & Angulo, I. (2016). State of the Art and Trends Review of Smart Metering in Electricity Grids. *Applied Sciences*, *6*(3), 68. doi:10.3390/app6030068

US Energy Independence and Security Act 2007 (EISA-2007)

Vader, N. V., & Bhadang, M. V. (2013). System integration: Smart grid with renewable energy. *Renewable Resources Journal*, *1*, 1–13.

Van Alstyne, M., Parker, G., & Choudary, P. (2016, April). Pipelines, Platforms, and the New Rules of Strategy. *Harvard Business Review*, 54-60, 62.

Van Geenhuisan, M., & Nijkamp, P. (1994). Sürdürülebilir kenti nasıl planlamalı? *Toplum ve Bilim Dergisi*, *64*(65), 129–140.

Vanolo, A. (2014). Smartmentality: The smart city as disciplinary strategy. *Urban Studies (Edinburgh, Scotland)*, *51*(5), 883–898. doi:10.1177/0042098013494427

VDMA. (n.d.). About the VDMA. *VDMA*. Retrieved from: http://www.vdma.org/

Viegas, J. L., Vieira, S. M., & Sousa, J. M. C. (2016). Mining consumer characteristics from smart metering data through fuzzy modelling. *Communications in Computer and Information Science*, *610*, 562–573. doi:10.1007/978-3-319-40596-4_47

Wang, W., He, Z., Huang, D., & Zhang, X. (2014). Research on Service Platform of Internet of Things for Smart City. In J. Jiang, & H. Zhang (Ed.), *ISPRS Technical Commission IV Symposium*. *XL-4* (pp. 301-303). Suzhou: Int. Arch. Photogramm. Remote Sens. Spatial Inf. Sci. 10.5194/isprsarchives-XL-4-301-2014

Wang, W., & Lu, Z. (2013). Cyber Security in the Smart Grid: Survey and Challenges. *Computer Networks*, *57*(5), 1344–1371. doi:10.1016/j.comnet.2012.12.017

Wang, Y., Hao, X., Song, L., Wu, C., Wang, Y., Hu, C., & Yu, L. (2012). Tracking states of massive electrical appliances by lightweight metering and sequence decoding(重点). *Proceedings of the Sixth International Workshop on Knowledge Discovery from Sensor Data*, 34–42. 10.1145/2350182.2350186

Warde, A. (1999). Convenient food: Space and timing. *British Food Journal*, *101*(7), 518–527. doi:10.1108/00070709910279018

Washburn, D., Sindhu, U., Balaouras, S., Dines, R. A., Hayes, N., & Nelson, L. E. (2009). Helping CIOs understand "smart city" initiatives. *Growth*, *17*(2), 1–17.

WCED (World Commission on Environment and Development). (1987). Our Common Future. Oxford, UK: WCED.

Weill, P., & Woerner, S. (2015). Thriving in an increasingly digital ecosystem. *MIT Sloan Management Review*, *56*(4), 27–34.

Weiser, M. (1991). The computer for the twenty-first century. *Scientific American*, *290*, 46–55.

Wenpeng, L., Sharp, D., & Lancashire, S. (2010). Smart grid communication network capacity planning for power utilities. *Proceedings of 2010 IEEE PES Transmission and Distribution Conference and Exposition*, 1-4. 10.1109/TDC.2010.5484223

Werner-Allen, G., Lorincz, K., & Marcillo, M. (2006). Deploying a Wireless Sensor Network on an Active Volcano. *IEEE Internet Computing*, 1-25.

Wijaya, T. K. b, Ganu, T., Chakraborty, D., Aberer, K., & Seetharam, D. P. (2014). Consumer segmentation and knowledge extraction from smart meter and survey data. In *SIAM International Conference on Data Mining 2014, SDM 2014* (*Vol. 1*, pp. 226–234). SIAM. 10.1137/1.9781611973440.26

Willness, C., & Bruni-Bossio, V. (2017). The curriculum innovation canvas: A design thinking framework for the engaged educational entrepreneur. *Journal of Higher Education Outreach & Engagement, 21*(1), 134–164.

Winter, M., Smith, C., Morris, P., & Cicmil, S. (2006). Directions for future research in project management: The main findings of UK government-funded research network. *International Journal of Project Management, 24*(8), 638–649. doi:10.1016/j.ijproman.2006.08.009

Woo, C., Jung, J., Euitack, J., Lee, J., Kwon, J., & Kim, D. (2016). Internet of Things Platform and Services for Connected Cars. In M. Ramachandran, G. Wills, R. Walters, V. Mendez Muñoz, & V. Chang (Ed.), *Proceedings of the International Conference on Internet of Things and Big Data* (pp. 469-478). Rome: Science and Technology Publications, Lda. 10.5220/0005952904690478

World Bank. (2017). Retrieved May 13, 2017, from http://www.worldbank.org

World Economic Forum. (2016). *Digital Transformation: Digital trends in the automotive industry*. Available at http://reports.weforum.org/digital-transformation/digital-trends-in the automotive-industry.Accessed 9/21/2017

Yang, Q. (2001). *Electrical engineering and its automation*. Beijing Sifang Automation Co. Ltd.

Yang, Y., Wu, L., & Hu, W. (2011). Security architecture and key technologies for power cloud computing. *Proc. IEEE Int. Conf. Transp., Mech., Electr. Eng.*, 1717–1720. 10.1109/TMEE.2011.6199543

Yang, Q., Barria, J. A., & Green, T. C. (2011). Communication infrastructures for distributed control of power distribution networks. *IEEE Transactions on Industrial Informatics*, *7*(2), 316–327. doi:10.1109/TII.2011.2123903

Yanliang, W., Song, D., Wei-Min, L., Tao, Z., & Yong, Y. (2010). Research of electric power information security protection on cloud security. *Proceeding IEEE International Conference on Power System Technology (POWERCON)*, 1–6.

Yan, Y., Qian, Y., Sharif, H., & Tipper, D. (2013). A Survey on Smart Grid Communication Infrastructures: Motivations, Requirements and Challenges. *IEEE Communications Surveys and Tutorials*, *15*(1), 5–20. doi:10.1109/SURV.2012.021312.00034

Yi, S., Hao, Z., Qin, Z., & Li, Q. (2015). *Fog Computing: Platform and Applications.* Paper presented at the Third IEEE Workshop on Hot Topics in Web Systems and Technologies. 10.1109/HotWeb.2015.22

Yigit, M., Gungor, V. C., Tuna, G., Rangoussi, M., & Fadel, E. (2014). Power line communication technologies for smart grid applications: A review of advances and challenges. *Computer Networks*, *70*, 366–383. doi:10.1016/j.comnet.2014.06.005

Yilmaz, E. (2016). *Yeni bir hikaye Konya akıllı şehir olabilir mi?* Retrieved May 30, 2016, from http://www.kto.org.tr/d/file/yeni-bir-hikaye-konya-akilli-sehir-olabilir-mi---emre-yilmaz.20160222152345.pdf

Yi, S., Li, C., & Li, Q. (2015). *A Survey of Fog Computing: Concepts, Applications and Issues. ACM Mobidata ('15).* Hangzhou, China: ACM. doi:10.1145/2757384.2757397

Young, D. (2008). When do energy-efficient appliances generate energy savings? Some evidence from Canada. *Energy Policy*, *36*(1), 34–46. doi:10.1016/j.enpol.2007.09.011

Yu, F., Zhang, F., Xiao, W., & Choudhury, P. (2011). Communication Systems for Grid Integration of Renewable Energy Resources. *IEEE Network*, *25*(5), 22–29. doi:10.1109/MNET.2011.6033032

Yu, Y., Yang, J., & Chen, B. (2012). The smart grids in China—A review. *Energies*, *5*(5), 1321–1338. doi:10.3390/en5051321

Zaballos, A., Vallejo, A., & Selga, J. (2011). Heterogeneous communication architecture for the smart grid. *IEEE Network*, *25*(5), 30–37. doi:10.1109/MNET.2011.6033033

Zala, H. N., & Abhyankar, R. (2014). A novel approach to design time of use tariff using load profiling and decomposition. *2014 IEEE International Conference on Power Electronics, Drives and Energy Systems (PEDES)*, 1–6. 10.1109/PEDES.2014.7042027

Zhang, X., Grijalva, S., & Reno, M. J. (2014). A time-variant load model based on smart meter data mining. *IEEE PES General Meeting / Conference & Exposition*, 1–5.

Zhang, Y., Wang, L., Sun, W., Green, R. II, & Alam, M. (2011). Distributed intrusion detection system in a multi-layer network architecture of smart grids. *IEEE Transactions on Smart Grid*, *2*(4), 796–808. doi:10.1109/TSG.2011.2159818

Zhao, J., Huang, W., Fang, Z., Chen, F., Li, K., & Deng, Y. (Eds.). (2007). *Proceedings of Power Engineering Society General Meeting*. doi:10.1109/PES.2007.385975

Zuboff, S. (1988). *In the Age of smart machines*. New York: Basic Books.

Бурматова, О. П. (2009). Управление воздействием отраслей экономики на окружающую среду: Учебное пособие [Managing the impact of industries on the environment: Textbook]. Novosibirsk: NSU.

В Японии официально открыт "умный город" Фудзисава. (2014). [In Japan, officially opened the "smart city" Fujisawa]. Retrieved from http://hitech.vesti.ru/news/view/id/6071

Государственные доклады «О состоянии и об охране окружающей среды Российской Федерации». (n.d.). [State reports "of the Russian Federation on the state and protection of the environment]. Retrieved from http://www.mnr.gov.ru/regulatory/list.php?part=1101

Государственный доклад "О состоянии и об охране окружающей среды Новосибирской области в 2015 году." (2016a). [The State Report "On the state and Environmental Protection of the Novosibirsk region in 2015]. Novosibirsk. Retrieved from http://dproos.nso.ru/sites/dproos.nso.ru/wodby_files/files/wiki/2014/12/gosdoklad-2015_0.pdf

Государственный доклад «О состоянии санитарно-эпидемиологического благополучия населения в Новосибирской области в 2015 году». (2016b). [The State Report "On the state sanitary and epidemiological welfare of the population in the Novosibirsk region in 2015"]. Novosibirsk. Retrieved from http://54.rospotrebnadzor.ru/c/document_library/get_file?uuid=ea420274-bb3f-4a95-aef6-620cb759f4fc&groupId=117057

Комбинированное (комплексное) действие ядов. (n.d.). [Combined (complex) action of poisons]. Retrieved from http://www.neonatology.narod.ru/toxicology/kompl_deistvie_jadov.html

Концепция охраны окружающей среды Новосибирской области на период до 2015 года. Утверждена распоряжением Губернатора Новосибирской области от 17.11.2009 N° 283-р. (n.d.). [The concept of environmental protection of the Novosibirsk region for the period up to 2015. Approved by order of the Governor of Novosibirsk region 17.11.2009 number 283-p.]. Retrieved from http://www.nso.ru/sites/test.new.nso.ru/wodby_files/files/migrate/activity/Socio-Economic_Policy/strat_plan /Documents/file895.pdf

Лысцов, В.Н., & Скотникова, О.Г. (1991). О возможности взаимного усиления вредных воздействий загрязняющих агентов окружающей среды [On the possibility of a mutually reinforcing harmful effects of polluting environmental agents]. *Журнал Всесоюзного Химического Общества им. Д.И. Менделеева, 1*, 61-65.

Львович Н.К. (2006). *Жизнь в мегаполисе* [Life in the megalopolis]. Nauka.

Навстречу «зеленой» экономике: пути к устойчивому развитию и искоренению бедности - обобщающий доклад для представителей властных структур. (2011). [Towards to a "green" economy: Pathways to sustainable development and poverty eradication - a synthesis report to representatives of authorities]. UNEP.

Петин, В.Г., & Сынзыныс, Б.И. (1998). *Комбинированное действие факторов окружающей среды на биологические системы* [The combined effect of environmental factors on biological systems]. Obninsk: IATE.

Программа реиндустриализации экономики Новосибирской области до 2025 года (утв. постановлением Правительства Новосибирской области от 01/04/2016 № 89-п). (2016). [Program of reindustrialization of the Novosibirsk Region's economy until 2025 (approved by the Government of the Novosibirsk region 04.01.2016 number 89-p)]. Retrieved from https://www.nso.ru/page/15755

Программа социально-экономического развития Новосибирской области до 2015 г. (n.d.). [The program of social and economic development of Novosibirsk region until 2015]. Retrieved from http://economy.newsib.ru/files/99713.pdf

Сибири, Э. (2009). стратегия и тактика модернизации [Economy of Siberia: The strategy and tactics of modernization]. Novosibirsk: Ankil.

Сонгдо — умный город будущего. (n.d.). [Songdo - smart city of the future]. Retrieved from http://green-agency.ru/songdo-umnyj-gorod-budushhego

Стратегический план устойчивого развития города Новосибирска и комплексные целевые программы. (2004). [Strategic Plan for Sustainable Development of the city of Novosibirsk and complex target programs]. Novosibirsk.

Стратегия социально-экономического развития Новосибирской области до 2025 года. (2007). [Strategy of Socioeconomic development of Novosibirsk oblast until 2025]. Novosibirsk. Retrieved from http://economnso.ru/files/1654.pdf

Умные города будущего: как строить полноценные интеллектуальные города в России. (n.d.). [The Smart Cities of the future: how to build a full-complete smart city in Russia]. Retrieved from http://sk.ru/utility/scripted-file.ashx?GroupKeys=net%2F1120292%2F&_cf=callback-standard-detail.vm&_fid=2609466&_ct=page&_cp=blogs-postlist&_ctt=c6108064af6511ddb074de1a56d89593&_ctc=1549&_ctn=7e987e474b714b01ba29b4336720c446&_cc=0&AppType=Weblog

Умные города: потенциал и перспективы развития в регионах России. Круглый стол (ГУ ВШЭ, Москва). (2014). [Smart Cities: Potential and prospects of development in the Russian regions. Round table. Retrieved from http://issuu.com/epliseckij/docs/bc9fac678b9405/5?e=7773934/7474790

Урбанизация и ее воздействие на состояние окружающей среды. (2011). [Urbanization and its impact on the environment]. Retrieved from http://phasad.ru/z1.php

Характеристики основных загрязнителей окружающей среды. (n.d.). [Characteristics of the main pollutants of the environment]. Retrieved from http://www.projects.uniyar.ac.ru/publish/ecostudy/toxic2.html#0

Чубик М.П. (2006). *Экология человека: Учебное пособие* [Human Ecology: Textbook]. Tomsk: Publishing House TSU.

Index

Recommended Reference Books

IGI Global's reference books can now be purchased from three unique pricing formats:
Print Only, E-Book Only, or Print + E-Book.
Shipping fees may apply.

www.igi-global.com

ISBN: 978-1-5225-9687-5
EISBN: 978-1-5225-9689-9
© 2020; 405 pp.
List Price: US$ 225

ISBN: 978-1-5225-9578-6
EISBN: 978-1-5225-9580-9
© 2019; 243 pp.
List Price: US$ 195

ISBN: 978-1-5225-5912-2
EISBN: 978-1-5225-5913-9
© 2019; 349 pp.
List Price: US$ 215

ISBN: 978-1-5225-3163-0
EISBN: 978-1-5225-3164-7
© 2018; 377 pp.
List Price: US$ 395

ISBN: 978-1-5225-2589-9
EISBN: 978-1-5225-2590-5
© 2018; 602 pp.
List Price: US$ 345

ISBN: 978-1-68318-016-6
EISBN: 978-1-5225-1989-8
© 2017; 197 pp.
List Price: US$ 180

Do you want to stay current on the latest research trends, product announcements, news, and special offers?
Join IGI Global's mailing list to receive customized recommendations, exclusive discounts, and more.
Sign up at: **www.igi-global.com/newsletters.**

Publisher of Peer-Reviewed, Timely, and Innovative Academic Research

www.igi-global.com　　Sign up at www.igi-global.com/newsletters　　f facebook.com/igiglobal　　t twitter.com/igiglobal

Ensure Quality Research is Introduced to the Academic Community

Become an Evaluator for IGI Global Authored Book Projects

The overall success of an authored book project is dependent on quality and timely manuscript evaluations.

Applications and Inquiries may be sent to:
development@igi-global.com

Applicants must have a doctorate (or equivalent degree) as well as publishing, research, and reviewing experience. Authored Book Evaluators are appointed for one-year terms and are expected to complete at least three evaluations per term. Upon successful completion of this term, evaluators can be considered for an additional term.

If you have a colleague that may be interested in this opportunity, we encourage you to share this information with them.

IGI Global Author Services

Providing a high-quality, affordable, and expeditious service, IGI Global's Author Services enable authors to streamline their publishing process, increase chance of acceptance, and adhere to IGI Global's publication standards.

Benefits of Author Services:

- **Professional Service:** All our editors, designers, and translators are experts in their field with years of experience and professional certifications.
- **Quality Guarantee & Certificate:** Each order is returned with a quality guarantee and certificate of professional completion.
- **Timeliness:** All editorial orders have a guaranteed return timeframe of 3-5 business days and translation orders are guaranteed in 7-10 business days.
- **Affordable Pricing:** IGI Global Author Services are competitively priced compared to other industry service providers.
- **APC Reimbursement:** IGI Global authors publishing Open Access (OA) will be able to deduct the cost of editing and other IGI Global author services from their OA APC publishing fee.

Author Services Offered:

English Language Copy Editing
Professional, native English language copy editors improve your manuscript's grammar, spelling, punctuation, terminology, semantics, consistency, flow, formatting, and more.

Scientific & Scholarly Editing
A Ph.D. level review for qualities such as originality and significance, interest to researchers, level of methodology and analysis, coverage of literature, organization, quality of writing, and strengths and weaknesses.

Figure, Table, Chart & Equation Conversions
Work with IGI Global's graphic designers before submission to enhance and design all figures and charts to IGI Global's specific standards for clarity.

Translation
Providing 70 language options, including Simplified and Traditional Chinese, Spanish, Arabic, German, French, and more.

Hear What the Experts Are Saying About IGI Global's Author Services

"Publishing with IGI Global has been *an amazing experience* for me for sharing my research. The *strong academic production* support ensures quality and timely completion." – **Prof. Margaret Niess, Oregon State University, USA**

"The service was *very fast, very thorough, and very helpful* in ensuring our chapter meets the criteria and requirements of the book's editors. I was *quite impressed and happy* with your service." – **Prof. Tom Brinthaupt, Middle Tennessee State University, USA**

Learn More or Get Started Here:
For Questions, Contact IGI Global's Customer Service Team at cust@igi-global.com or 717-533-8845

www.igi-global.com

www.igi-global.com

Celebrating Over 30 Years of Scholarly
Knowledge Creation & Dissemination

InfoSci®-Books

A Database of Nearly 6,000 Reference Books Containing Over 105,000+ Chapters Focusing on Emerging Research

GAIN ACCESS TO **THOUSANDS** OF
REFERENCE BOOKS AT **A FRACTION**
OF THEIR INDIVIDUAL LIST **PRICE**.

InfoSci®-Books Database

The **InfoSci®-Books** is a database of nearly 6,000 IGI Global single and multi-volume reference books, handbooks of research, and encyclopedias, encompassing groundbreaking research from prominent experts worldwide that spans over 350+ topics in 11 core subject areas including business, computer science, education, science and engineering, social sciences, and more.

Open Access Fee Waiver (Read & Publish) Initiative

For any library that invests in IGI Global's InfoSci-Books and/or InfoSci-Journals (175+ scholarly journals) databases, IGI Global will match the library's investment with a fund of equal value to go toward **subsidizing the OA article processing charges (APCs) for their students, faculty, and staff** at that institution when their work is submitted and accepted under OA into an IGI Global journal.*

INFOSCI® PLATFORM FEATURES

- Unlimited Simultaneous Access
- No DRM
- No Set-Up or Maintenance Fees
- A Guarantee of No More Than a 5% Annual Increase for Subscriptions
- Full-Text HTML and PDF Viewing Options
- Downloadable MARC Records
- COUNTER 5 Compliant Reports
- Formatted Citations With Ability to Export to RefWorks and EasyBib
- No Embargo of Content (Research is Available Months in Advance of the Print Release)

*The fund will be offered on an annual basis and expire at the end of the subscription period. The fund would renew as the subscription is renewed for each year thereafter. The open access fees will be waived after the student, faculty, or staff's paper has been vetted and accepted into an IGI Global journal and the fund can only be used toward publishing OA in an IGI Global journal. Libraries in developing countries will have the match on their investment doubled.

To Recommend or Request a Free Trial:
www.igi-global.com/infosci-books

eresources@igi-global.com • Toll Free: 1-866-342-6657 ext. 100 • Phone: 717-533-8845 x100

www.igi-global.com

www.igi-global.com

Publisher of Peer-Reviewed, Timely, and
Innovative Academic Research Since 1988

IGI Global's Transformative Open Access (OA) Model:
How to Turn Your University Library's Database Acquisitions Into a Source of OA Funding

Well in advance of Plan S, IGI Global unveiled their OA Fee
Waiver (Read & Publish) Initiative. Under this initiative, librarians
who invest in IGI Global's InfoSci-Books and/or InfoSci-Journals
databases will be able to subsidize their patrons' OA article
processing charges (APCs) when their work is submitted and
accepted (after the peer review process) into an IGI Global journal.

How Does it Work?

Step 1: **Library Invests in the InfoSci-Databases:** A library perpetually purchases or subscribes to the InfoSci-Books, InfoSci-Journals, or discipline/subject databases.

Step 2: **IGI Global Matches the Library Investment with OA Subsidies Fund:** IGI Global provides a fund to go towards subsidizing the OA APCs for the library's patrons.

Step 3: **Patron of the Library is Accepted into IGI Global Journal (After Peer Review):** When a patron's paper is accepted into an IGI Global journal, they option to have their paper published under a traditional publishing model or as OA.

Step 4: **IGI Global Will Deduct APC Cost from OA Subsidies Fund:** If the author decides to publish under OA, the OA APC fee will be deducted from the OA subsidies fund.

Step 5: **Author's Work Becomes Freely Available:** The patron's work will be freely available under CC BY copyright license, enabling them to share it freely with the academic community.

Note: *This fund will be offered on an annual basis and will renew as the subscription is renewed for each year thereafter. IGI Global will manage the fund and award the APC waivers unless the librarian has a preference as to how the funds should be managed.*

Hear From the Experts on This Initiative:

"I'm very happy to have been able to make one of my recent research contributions *freely available* along with having access to the *valuable resources* found within IGI Global's InfoSci-Journals database."

– **Prof. Stuart Palmer**,
Deakin University, Australia

"Receiving the support from IGI Global's OA Fee Waiver Initiative *encourages me to continue my research work without any hesitation*."

– **Prof. Wenlong Liu**, College of
Economics and Management at
Nanjing University of Aeronautics &
Astronautics, China

For More Information, Scan the QR Code or Contact:
IGI Global's Digital Resources Team at eresources@igi-global.com.

Printed in the United States
by Baker & Taylor Publisher Services